Thesaurus of Information Science Terminology

Completely revised and expanded edition

by

CLAIRE K. SCHULTZ

The Scarecrow Press, Inc.
Metuchen, N.J. & London
1978

Library of Congress Cataloging in Publication Data

Schultz, Claire K
Thesaurus of information science terminology.

Includes index.
1. Subject headings--Information science.
I. Title.
Z695.1.I56S36 1978 025.33'02 78-16878
ISBN 0-8108-1156-1

Manufactured in the United States of America

TABLE OF CONTENTS

PREFACE

This thesaurus treats the realm of information science as a system. Therefore, its taxonomy is based on five major divisions, namely: environment, input, processing, output, and feedback & control, in keeping with the cybernetic systems theory. Within each of these five contexts, traditional hierarchic structures are developed to the best of this compiler's ability, given her understanding of information science and the field's current state of development.

To most thesaurus users, the basis of the taxonomy will be of no great concern as long as order of some kind can be demonstrated and the method of ordering is found workable in practical situations. Previous editions of this thesaurus did not succeed in developing a strong ordering of the terminology, probably because it was less clear then than now what the scope and interrelations of information science are. Hopefully, this edition of the thesaurus demonstrates evolving insights within the field of information science concerning both its boundaries and its intra- and interdisciplinary relationships.

Limitations of This Thesaurus

In former editions, this author employed the caveat that the thesaurus was developed empirically for a private collection. The caveat still holds; however, the collection has gradually become more voluminous and somewhat broader in scope. At this writing, that collection (mostly government research reports and journal articles) numbers between four and five thousand documents.

Subject matter of the collection still centers on storage and retrieval, and automation of the same; but there are other segments of information science represented in enough detail to at least sketch the field as a whole.

Some of the limitations of this thesaurus are declared here, not only to free the compiler from the burdens of perfection, but to make clear that persons adapting this thesaurus to their own needs are expected to view it as only a stepping stone for meeting their requirements. Some persons, such as students in a library school indexing course may be able to use this thesaurus just as it is and it is because of the student audience that it seems worthwhile to publish it. Those who need to revise it will be helped by whatever portions of its basic development they can use, since establishing a sound framework is one of the hardest parts of the task.

A Tutorial Section Accompanies This Thesaurus

Instructions for how to construct, use and revise a thesaurus are given in a separate tutorial section of this book. The tutorial chapters were written for the benefit of students but they are there for the use of other uninitiated persons as well.

Continuity With Previous Editions

To a large extent this thesaurus is compatible with previous editions, but some of its terminology is new and certain terms have been differently defined or split into more than one term. Persons who have retrieval systems based on a previous edition may want to do term by term comparisons before deciding whether to accept the new edition or modify the old edition to suit their purposes.

ACKNOWLEDGMENTS

Three women, besides the compiler, labored to produce this book.

Rita Porreca, systems programmer, and esteemed colleague for more than a decade, wrote the computer programs needed to do much of the editorial construction work and also much of the formatting of the thesaurus sections for automated printing.

Hilary Hoguet, data processing coordinator, worked in concert with the computer programs from a remote terminal, patiently feeding the computer corrected data as the compiler iteratively revised the content. She used a sophisticated formatting program to insert running heads, page numbers, discretionary hyphens, and other typesetting requirements, to produce well-formed pages.

Barbara Gordon, data processing clerk, keyed the front matter and the tutorial sections of this book, working out the format as she proceeded, until revision was compete and a master copy was produced for the publisher.

The compiler is greatly indebted to all three women.

HOW TO USE THIS THESAURUS

Note: It is suggested that persons unfamiliar with use of thesauri for storage and retrieval purposes consult at least Chapter 1 of the tutorial section of this book to learn the basic functions of thesauri.

Alphabetic Arrangement of the Thesaurus

The alphabetic arrangement of the thesaurus has two types of entry:

1) "Unaccepted" or "use" terms,
   type A: (term X use term Y)
   type B: (term X use terms Y and Z).

2) "Accepted" terms, those terms which are provided for use during indexing or retrieval.

"Accepted" terms are accompanied by one or more types of note, meant to help the user understand the term and its relations with other entries. Briefly, they are, in order of the appearance within an entry:

> Scope note: A scope note can either be labeled "Note" as in the entry:
>
> INFORMATION
> Note: Use more specific or precise terms instead whenever possible; see the related terms

for examples.

or it can be a parenthetic expression which is actually part of accepted terms as in:

AGE (OF PERSONS OR MATERIAL).
The purpose of the scope note is to help to define how the term is used within the context of the thesaurus.

X: Refer from: This note gathers terms which "use" entries of type A generate. Considering the example (X use Y), Y would note X in its "refer from:" note.

BT: Broader terms: The term that is hierarchically one level superior to the entry term in the structured arrangement of the thesaurus.

NT: Narrower terms: The terms which are hierarchically one level inferior to the entry term in the structured arrangement of the thesaurus.

RT: Related terms: Terms either closely related in the hierarchic structure, or terms which the user may want to consider for other semantic reasons. Related terms are reciprocal; that is, if term A notes term B as a related term, then B notes A as a related term, also.

PX: Partial refer from: This note gathers elements from "use" terms of type B. Considering the example (X use Y and Z), Y would note X as a "partial refer from" and Z would also note X as a "partial refer from."

The thesaurus user is exhorted to make full use of the notes accompanying any term being consulted, both to save time and to benefit from the contextual information which they present.

Many terms are not accompanied by a full set of notes because, for example, a term may not have any hierarchically inferior terms, or may not have been referenced in any "use" entries. "Related terms" are generated by the compiler as a potential convenience to the user and are not to be relied upon for exhaustiveness. Depending on the view of the particular user, some of them will not be semanticallly interesting and should be disregarded in such cases.

It is probably obvious that if the user wishes to revise a thesaurus entry, the revision will affect other terms as well. For example, if an entry term is deleted or changed, then every term specified within its system of notes must be changed in conformity. Thus, the notes have a housekeeping function, in addition to being aids to effective storage and retrieval.

Hierarchic Arrangement of the Thesaurus

Every "accepted" term has a unique place in the hierarchy, of which "information science" is the top node. As was explained in the preface, the nodes directly under "information science" divide information science into standard cybernetic system aspects, which can be depicted as:

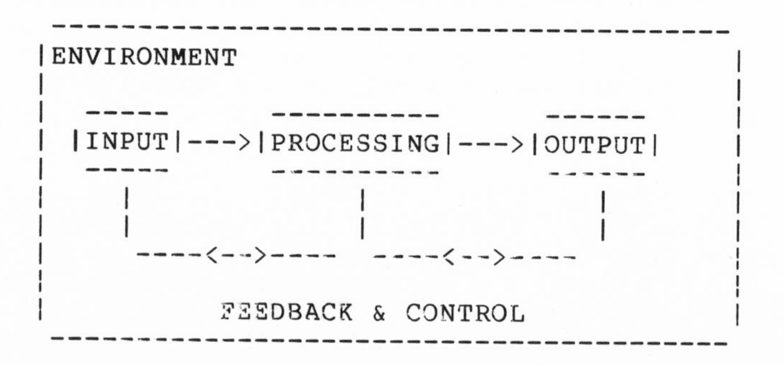

The compiler attempted to separate terms according to whether they constituted the environment of information science, its products, or its processes, and so on, as an interesting intellectual exercise and as one meant to help define both the scope of the field and what it does. Most terms could occur in relationships other than those depicted in the hierarchy. It has always been the plight of taxonomists that two dimensions are not sufficient for capturing the essence of most concepts or entities, no matter how one goes about it.

The thesaurus user is asked to use the hierarchy for what it is worth, that is, as one form of ordering which ties all of the content of the thesaurus into a conceptual whole. However, the terms can be coordinated with one another singly as in coordinate indexing:

Abstracting
Ability
Documents
Science and Technology

or in pre-coordinated indexing:

Ability - Abstracting
Documents - Science and Technology.

ALPHABETIC ARRANGEMENT OF THE THESAURUS

A

AACR (Anglo-American Cataloging Rules)
  use INDEXING
      BIBLIOGRAPHY (PROCESS)

ABA (American Bar Association)
  use LAW
      PROFESSIONAL ORGANIZATIONS

Abbreviations
  use CODING SYSTEMS

ABILITY
  X Aptitude
     Leadership
  BT Human Resources
  NT Creativity
     Skills
  PX Illiteracy
     Literacy

Abridgement (Abridging)
  use EXTRACTING

Abstracter
  use ABSTRACTING
      INFORMATION SCIENCE PERSONNEL

ABSTRACTING
  BT Bibliography (Process)
  RT Abstracts (Of Documents)
     Evaluation
     Evaluation Techniques

Extracting
Publications (Personal Or Organizational)
Service Organizations
PX Abstracter
Abstracting Services
Author Abstracting
Autoabstracting
National Federation Of Abstracting And Indexing Services

Abstracting Services
use ABSTRACTING
INFORMATION SCIENCE SERVICE ORGANIZATIONS

ABSTRACTS (OF DOCUMENTS)
X Condensations (Of Literature)
Extracts
Resumes (Of Documents)
Summaries
BT Documents
NT Abstracts Publications
RT Abstracting
Indicators
Reviews

ABSTRACTS PUBLICATIONS
BT Abstracts (Of Documents)
PX Biological Abstracts (Publication)
Chemical Abstracts (Publication)
Excerpta Medica

Academic Libraries
use ACADEMIC ORGANIZATIONS
LIBRARIES

ACADEMIC ORGANIZATIONS
X Colleges
Educational Institutions
Schools
Universities
BT Organizations
RT Consulting Services
Education
Libraries
PX Academic Libraries
Academic Status
Degrees (Academic)
Educom (Educational Communications)

Honorary Degrees
Inter-University Communications Council
Library Schools
Research Libraries
University Libraries

Academic Status
use ACADEMIC ORGANIZATIONS
AWARDS

Accessibility
use AVAILABILITY

ACCESSING
X Entry (To Gain Access)
BT Interaction
PX DDD (Direct Distance Dialing)
Dial-Access Information Retrieval
Direct Distance Dialing (DDD)
Random Access

Accession (Adding To)
use ACQUIRING

Accountability
use RESPONSIBILITY

ACCOUNTING
X Inventorying
BT Finance
RT Budgeting
Electromechanical Data Processing Equipment
Electronic Digital Computers
PX Cost Accounting
Cost Analysis

Accreditation
use AWARDS

ACLU (American Civil Liberties Union)
use RIGHTS
SERVICE ORGANIZATIONS

ACM (Association For Computing Machinery)
use DATA PROCESSING
INFORMATION SCIENCE PROFESSIONAL ORGANIZATIONS

Acoustic Couplers
  use INPUT-OUTPUT EQUIPMENT

Acoustics
  use SOUND

ACQUIRING
  X   Accession (Adding To)
       Collecting
       Order (Purchase)
       Procurement (Of Documents)
       Purchasing
  BT   Input Process
  RT   Budgeting
       Control Functions
       Costs
       Data Gathering
  PX   Information Collecting (Documents)

Acronym
  use CODING SYSTEMS

ACS (American Chemical Society)
  use CHEMISTRY
       SERVICE ORGANIZATIONS

Adapting
  use CHANGE

Adaptive Computer Systems
  use MAN-MACHINE INTERACTION

ADBS (Assoc. Franc. Des Documentalistes Et Des
     Bibliothecaires Specialises)
  use INFORMATION SCIENCE PROFESSIONAL
         ORGANIZATIONS
       FRANCE

Added Entry
  use INDEX TERMS (FORMAT AND INTERRELATIONS)

Address Lists
  use NAME AND ADDRESS LISTS

Adequacy
  use EFFECTIVENESS

ADI (American Documentation Institute)

use INFORMATION SCIENCE PROFESSIONAL ORGANIZATIONS

ADJUSTMENT
BT Interaction

ADMINISTRATION
X Directing
Management
Supervision
BT Feedback And Control
NT Control Functions
Employment
Finance
Interaction
Interrelations
Uses (Of Resources)
RT Criteria
Evaluation
Responsibility
PX Administrators
AMA (American Management Association)
American Management Association (AMA)
Boards (Of Governance)
Business Administration
Commissions (Administrative Groups)
Coordinators
Executives
Library Administration
Resource Management

Administrators
use ADMINISTRATION
PERSONNEL

Adult Education
use ADULTS
LEARNING

ADULTS
BT Roles (Social)
NT Aged
Young Adults
RT Children
PX Adult Education

Advanced Research Projects Agency (ARPA)
use INFORMATION SCIENCE

RESEARCH

ADVERTISING
  BT Marketing
  RT Consumers
     Documents
     Exhibiting
     Finance

ADVERTISING LITERATURE
  BT Documents
  RT Exhibiting
     Lay Publications
     News Publications

Aeronautics
  use AEROSPACE SCIENCES

AEROSPACE SCIENCES
  X Aeronautics
    Space (Outer Space)
  BT Physical Sciences

AFIPS (American Federation Of Information Processing Societies)
  use DATA PROCESSING
      INFORMATION SCIENCE PROFESSIONAL ORGANIZATIONS

AFRICA
  BT Geographic Areas

AGE (OF PERSONS OR MATERIAL)
  BT Time

AGED
  X Senior Citizens
  BT Adults

Agencies
  use ORGANIZATIONS

Agenda
  use PLANNING

Agenda (Document)
  use MEETING DOCUMENTS

AGRICULTURE
  BT   Biological Sciences

AIBS (American Institute Of Biological Sciences)
  use   BIOLOGICAL SCIENCES
        PROFESSIONAL ORGANIZATIONS

Aides
  use   PERSONNEL
        PARAPROFESSIONALS

AIM (Abridged Index Medicus)
  use   BIOMEDICAL SCIENCES
        INDEXES

AIP (American Institute Of Physics)
  use   PHYSICAL SCIENCES
        PROFESSIONAL ORGANIZATIONS

Air Pollution Information Center (APTIC)
  use   ECOLOGY

ALA (American Library Association)
  use   LIBRARY SCIENCE
        INFORMATION SCIENCE PROFESSIONAL
            ORGANIZATIONS

ALGEBRAS
  X   Maps (Algebraic)
  BT   Mathematics
  NT   Boolean Algebra
  RT   Arrangement
       Graphic Records
       Logic
       Modeling
  PX   Lattices

ALGOL (PROGRAMMING LANGUAGE)
  BT   Artificial Languages

Algorithms
  use   PROGRAMS

Allocation
  use   PLANNING

Alphabetic Codes
  use   CODING SYSTEMS

Alphabetic Subject Code
  use INDEXES

ALPHABETS
  BT Symbol Sets
  NT Braille

Alphanumeric Codes
  use CODING SYSTEMS

AMA (American Management Association)
  use ADMINISTRATION
    PROFESSIONAL ORGANIZATIONS

AMA (American Medical Association)
  use BIOMEDICAL SCIENCES
    PROFESSIONAL ORGANIZATIONS

Amalgamate
  use MERGING

American Bar Association (ABA)
  use LAW
    PROFESSIONAL ORGANIZATIONS

American Chemical Society (ACS)
  use CHEMISTRY
    SERVICE ORGANIZATIONS

American Civil Liberties Union (ACLU)
  use RIGHTS
    SERVICE ORGANIZATIONS

American Documentation (Journal)
  use INFORMATION SCIENCE
    PERIODICALS

American Documentation Institute (ADI)
  use INFORMATION SCIENCE PROFESSIONAL
      ORGANIZATIONS

American Federation Of Information
    Processing Socieites (AFIPS)
  use DATA PROCESSING
    INFORMATION SCIENCE PROFESSIONAL
      ORGANIZATIONS

American Institute Of Biological Sciences (AIBS)

use BIOLOGICAL SCIENCES
PROFESSIONAL ORGANIZATIONS

American Institute Of Physics (AIP)
use PHYSICAL SCIENCES
PROFESSIONAL ORGANIZATIONS

American Library Association (ALA)
use LIBRARY SCIENCE
INFORMATION SCIENCE PROFESSIONAL ORGANIZATIONS

American Management Association (AMA)
use ADMINISTRATION
PROFESSIONAL ORGANIZATIONS

American Mathematical Society (AMS)
use MATHEMATICS
PROFESSIONAL ORGANIZATIONS

American Medical Association (AMA)
use BIOMEDICAL SCIENCES
PROFESSIONAL ORGANIZATIONS

American National Standards Institute (ANSI)
use SERVICE ORGANIZATIONS
SPECIFICATIONS

American Petroleum Institute (API)
use SERVICE ORGANIZATIONS

American Psychological Association (APA)
use BEHAVIORAL SCIENCES
PROFESSIONAL ORGANIZATIONS

American Society For Information Science (ASIS)
use INFORMATION SCIENCE PROFESSIONAL ORGANIZATIONS

American Society For Metals (ASM)
use METALLURGY
PROFESSIONAL ORGANIZATIONS

American Standard Code For Information Interchange (ASCII)
use DATA PROCESSING
CODING SYSTEMS

American Standards Organization
  use SERVICE ORGANIZATIONS
      SPECIFICATIONS

AMS (American Mathematical Society)
  use MATHEMATICS
      PROFESSIONAL ORGANIZATIONS

ANALOG EQUIPMENT
  BT Computers
  RT Automata
      Electronic Digital Computers
      Hybrid Computers

Analog-Digital Conversion
  use DATA
      TRANSLATION

Analog-Digital Conversion Equipment
  use TRANSLATION
      EQUIPMENT

Analysis, Data
  use DATA ANALYSIS

Analysis, Document
  use INDEXING

Analysis, Information
  use DATA ANALYSIS

Analysis, Linguistic
  use LINGUISTICS

Analysis, Mathematical
  use MATHEMATICS

Analysis, Statistical
  use DATA ANALYSIS

Analysis, Syntactic
  use SYNTAX

Analytical Subject Entry
  use INDEXING

Analyzing
  use EVALUATION

Anglo-American Cataloging Rules (AACR)
  use INDEXING
    BIBLIOGRAPHY (PROCESS)

ANIMALS
  BT Population

Annotated Bibliographies
  use BIBLIOGRAPHIES
    REVIEWS

Annotating
  use REVIEWS

Announcement Media
  use NEWS PUBLICATIONS

Announcement Services
  use DISSEMINATING

Annual Review Of Information Science
    And Technology (ARIST)
  use INFORMATION SCIENCE
    REVIEWS

Annual Reviews
  use REVIEWS

ANSI (American National Standards Institute)
  use SERVICE ORGANIZATIONS
    SPECIFICATIONS

ANTHROPOLOGY
  BT Behavioral Sciences
  RT Cultures

APA (American Psychological Association)
  use BEHAVIORAL SCIENCES
    PROFESSIONAL ORGANIZATIONS

API (American Petroleum Institute)
  use SERVICE ORGANIZATIONS
    SCIENCE AND TECHNOLOGY

APL (PROGRAMMING LANGUAGE)
  BT Artificial Languages

APPLICANTS

BT Roles (Social)

APPLICATIONS
Note: Used For Applying Both Equipment And Techniques.
X Projects
BT Uses (Of Resources)
NT Computer Applications
Operation (Of Equipment)
RT Computers
Designing
Equipment
Techniques
PX Batten System
Classification (Process)
Coding Fields
Criticism
Data Processing Systems
Microfilm Systems
Program (Of Activities)
Punched Card Systems
Standardization
Stimulation
System Analysis
System Design

Applications, Computer
use COMPUTER APPLICATIONS

Applied Linguistics
use LINGUISTICS

Approach
use TECHNIQUES

APS (Auxiliary Publication Service)
use PUBLISHING
FILM (PHOTOGRAPHIC)
INFORMATION SCIENCE SERVICE ORGANIZATIONS

APTIC (Air Pollution Information Center)
use ECOLOGY
INFORMATION SCIENCE SERVICE ORGANIZATIONS

Aptitude
use ABILITY

ARABIC LANGUAGE

BT Natural Languages

ARCHITECTURE
BT Fine Arts
RT Buildings

ARCHIVAL RESEARCH
X Historical Research
BT Research

Archives (Organized)
use LIBRARIES

ARIST (Annual Review Of Information Science And Technology
use INFORMATION SCIENCE
REVIEWS

ARL (Association Of Research Libraries)
use LIBRARY SCIENCE
INFORMATION SCIENCE PROFESSIONAL ORGANIZATIONS

ARPA (Advanced Research Projects Agency)
use INFORMATION SCIENCE
RESEARCH

Arpanet
use TIME-SHARING SERVICES
INFORMATION SCIENCE
RESEARCH

ARRANGEMENT
X Forms
Matrices
BT Structure
NT Classification Systems
RT Algebras
Coding Systems
Numbering Systems
PX Articulated Indexes
Classified Index
Continuous Records
Conventional Systems (Of Indexing)
Discontinuous Records
Keyword-Out-Of-Context Indexes
KWOC Indexes
Microfiche (Medium)

Unit Records
Vertical Files

Arranging
use INTERRELATIONS

Articles (Parts Of Speech)
use SYNTAX

Articulated Indexes
use INDEXES
ARRANGEMENT

Artificial Intelligence
use INTELLECT (INTELLIGENCE)
AUTOMATA

ARTIFICIAL LANGUAGES
X Compiler Languages
Intermediate Languages
Machine Languages
Programming Languages
BT Languages
NT ALGOL (Programming Language)
APL (Programming Language)
BASIC (Programming Language)
COBOL (Programming Language)
COURSE WRITER (Programming Language)
FORTRAN (Programming Language)
Interlingua
LISP (Programming Language)
MUMPS (Programming Language)
PL/1 (Programming Language)
PX Data Description Languages
String-processing Languages

ARTS AND HUMANITIES
X Drama
Humanities
BT Knowledge Areas
NT Fine Arts
RT Dancing

ASCA (Automatic Subject Citation Alert) (Tradename)
use INFORMATION FLOW
TECHNIQUES

ASCII (American Standard Code For Information Interchange)
use DATA PROCESSING
CODING SYSTEMS

ASIA
BT Geographic Areas
NT China
India
Israel
Japan
Korea
Pakistan

ASIS (American Society For Information Science)
use INFORMATION SCIENCE PROFESSIONAL ORGANIZATIONS

ASLIB (Association Of Library And Information Bureaux)
use INFORMATION SCIENCE PROFESSIONAL ORGANIZATIONS

ASM (American Society For Metals)
use METALLURGY
PROFESSIONAL ORGANIZATIONS

Aspect (Indexing Term)
use INDEX TERMS (DOCUMENT SURROGATES)

Assembling
use CONSTRUCTING

Assembling (Computer Programs)
use TRANSLATION
PROGRAMS

Assessing
use REVIEWS

Assistants
use PERSONNEL

Assoc. Franc. Des Documentalistes Et Des Bibliothecaires Specialises (ADBS)
use INFORMATION SCIENCE PROFESSIONAL ORGANIZATIONS
FRANCE

ASSOCIATING
  BT   Interrelations
  NT   Browsing
  RT   Correlating
       Intellect (Intelligence)
  PX   Associative Memory
       Content-Addressable Memories

Association For Computer Machinery (ACM)
  use DATA PROCESSING
       INFORMATION SCIENCE PROFESSIONAL
           ORGANIZATIONS

Association Of Library And Information Bureaux
       (ASLIB)
  use INFORMATION SCIENCE PROFESSIONAL
           ORGANIZATIONS
       UNITED KINGDOM

Association Of Research Libraries (ARL)
  use LIBRARY SCIENCE
       INFORMATION SCIENCE PROFESSIONAL
           ORGANIZATIONS

Associations
  use ORGANIZATIONS

Associative Memory
  use ASSOCIATING
       MEMORY

ATOMIC AND MOLECULAR PHYSICS
  X   Atomic Energy
  BT   Physical Sciences

Atomic Energy
  use ATOMIC AND MOLECULAR PHYSICS

Attendance Lists
  use MEETING DOCUMENTS

Attendants
  use PERSONNEL

ATTITUDES
  X   Commitment
       Morale
       Opinions
       Outlook

BT Personality
RT Expectations
Motivation
PX Editorials

Attorneys
use LAW
PERSONNEL

Attribute (Indexing Term)
use INDEX TERMS (DOCUMENT SURROGATES)

Audio Cassettes
use SOUND
MAGNETIC TAPE

Audio Input-Output
use SOUND
INPUT-OUTPUT EQUIPMENT

Audiovisual Centers
use INFORMATION SCIENCE SERVICE ORGANIZATIONS

AUDIOVISUAL EQUIPMENT
Note: Use Only When Combinations Of More Specific Terms Such As "Equipment; Oral Communication" Or "Equipment; Visual Communication" Cannot Be Used Instead.
X Phonograph
Playback Systems
BT Input-Output Equipment
NT Television Equipment
RT Oral Communication

AUDIOVISUAL PRODUCTION SERVICES
BT Delivery Of Services
RT Graphic Records
Printing Services
Publishing Services

AUDIOVISUAL SOFTWARE (DOCUMENTS)
X Nonbook Materials
BT Documents
NT Graphic Records
Microform Documents
Motion Pictures

Music (Documents)
Slides
Sound Recordings
Videotapes (Documents)
RT Multimedia Techniques
Oral Communication
Records
Sound
Visual Communication

Audiovisual Software (Media)
use STORAGE MEDIA

AUSTRALIA
BT Geographic Areas

Author Abstracting
use AUTHORS
ABSTRACTING

Author Indexing
use AUTHORS
INDEXING

Authority
use CONTROL FUNCTIONS

Authority Lists
use THESAURI

AUTHORS
BT Roles (Social)
RT Publishing
PX Author Abstracting
Author Indexing
Source Indexing

Autoabstracting
use ABSTRACTING
COMPUTER APPLICATIONS

Autobiographies
use BIOGRAPHICAL DOCUMENTS

Autocorrelation
use CORRELATING

AUTOMATA

Note: Used Only When More Specific Terms Such as Computers Cannot Be Used Instead.
BT Equipment
RT Analog Equipment
Electronic Digital Computers
Hybrid Computers
PX Artificial Intelligence

Automated Coding
use CODING
COMPUTER APPLICATIONS

Automated Indexing
use INDEXING
COMPUTER APPLICATIONS

Automated Information Processing
use DATA PROCESSING

Automated Language Processing
use LINGUISTICS
COMPUTER APPLICATIONS

Automated Language Translation
use LANGUAGE TRANSLATION
COMPUTER APPLICATIONS

Automated Law Enforcement
use LAW
COMPUTER APPLICATIONS

Automated Medical Diagnosis
use DIAGNOSIS
COMPUTER APPLICATIONS

Automated Medical History Systems
use MEDICAL RECORDS
COMPUTER APPLICATIONS

Automated Reading
use READING
INPUT-OUTPUT EQUIPMENT

Automated Retrieval
use STORAGE AND RETRIEVAL PROCESSES
COMPUTER APPLICATIONS

Automated Speech
  use SPEECH
      INPUT-OUTPUT EQUIPMENT

Automated Subject Analysis
  use INDEXING
      COMPUTER APPLICATIONS

Automated Teaching
  use INSTRUCTING
      MAN-MACHINE INTERACTION

Automated Translation (Transformation)
  use TRANSLATION
      ELECTRONIC DIGITAL COMPUTERS

Automated Typesetting
  use PRINTING
      COMPUTER APPLICATIONS

Automatic Subject Citation Alert (ASCA)
    (Tradename)
  use INFORMATION FLOW
      TECHNIQUES

Automation
  use COMPUTER APPLICATIONS

Auxiliary Publication Service (APS)
  use FILM (PHOTOGRAPHIC)
      PUBLISHING
      INFORMATION SCIENCE SERVICE ORGANIZATIONS

AVAILABILITY
  X Accessibility
  BT Interaction
  RT Browsing
     Readiness

AWARDS
  X Accreditation
    Certification
    Election (To Office)
    Scholarships
  BT Compensation
  PX Academic Status
     Degrees (Academic)
     Honorary Degrees

B

Back-Of-The-Book Indexes
  use INDEXES
      BOOKS

BALLOTS (Bibliographic Automation Of Large Libraries Using Time-Sharing)
  use LIBRARIES
      TIME-SHARING SERVICES

BANKING
  BT Economics

BARRIERS
  X Obstacles
  BT Relations
  PX Barriers, Language
      Language Barriers

Barriers, Language
  use LANGUAGES
      BARRIERS

BASIC (PROGRAMMING LANGUAGE)
  BT Artificial Languages

BATCH PROCESSING
  BT Data Processing

Batch Programming
  use PROGRAMMING

Batten System
  use OPTICAL COINCIDENCE TECHNIQUES

BCN (Biomedical Communications Network)
  use BIOMEDICAL SCIENCES
      TIME-SHARING SERVICES

BEHAVIOR
  X Human Behavior
      Life Style
      Misdemeanors
      Passivity
      Personal Adjustment
      Smoking (Behavior)

BT Psychology
NT Conditioning
Performance
Recreation
Resistance (Human)
Responsibility
RT Human Relations
Linguistics
Personnel
PX Communication Behavior
Group Behavior
Information Seeking Behavior
Vandalism

BEHAVIORAL SCIENCES
BT Science And Technology
NT Anthropology
Economics
Education
History
Law
Library Science
Linguistics
Military Science
Political Science
Psychology
RT Biomedical Sciences
PX American Psychological Association (APA)
APA (American Psychological Association)
Social Scientists

BENEFIT
BT Values
RT Designing
Effectiveness
Efficiency
Feasibility
Measures (Measuring Techniques)
PX Cost Benefit

BIAS
X Prejudice
BT Interrelations
RT Weighting

Bibliographic Centers
use INFORMATION SCIENCE SERVICE ORGANIZATIONS

Bibliographic Control
  use BIBLIOGRAPHY (PROCESS)

Bibliographic Coupling
  use CITATION INDEXES

Bibliographic Data
  use CITATIONS

Bibliographic Data Bases
  use INDEXES
      TIME-SHARING SERVICES

Bibliographic Elements
  use CITATIONS

Bibliographic On-Line Display (BOLD) (Tradename)
  use TIME-SHARING SERVICES

Bibliographic Search
  use SEARCHING

BIBLIOGRAPHIES
  X Citation, Literature
    Citations (Bibliographies)
  BT Documents
  NT Citations
  RT Citation Indexes
    Data Processing Services
    Indexes
    Responses (To Inquiries)
  PX Annotated Bibliographies
    Catalogs
    Citation Studies

BIBLIOGRAPHY (PROCESS)
  X Bibliographic Control
    Citation (Process)
    Compiling (Bibliographies)
    Information Management
    Secondary Publications (Indexes; Abstracts
      Publications; Reviews)
  BT Library Science
  NT Abstracting
    Indexing
  RT Citations
    Control Functions
    Information

PX AACR (Anglo-American Cataloging Rules)
Anglo-American Cataloging Rules (AACR)
BNB (British National Bibliography)
British National Bibliography (BNB)
Descriptive Cataloging (Process)
MARC (Machine-readable Cataloging)

Bibliotherapy
use READING

Bidding
use PROPOSALS

Binary Codes
use CODING SYSTEMS

Binary Numbering
use NUMBERING SYSTEMS

BINDING (OF DOCUMENTS)
use BINDING SERVICES

BINDING SERVICES
BT Delivery Of Services
X Binding (Of Documents)

Biochemistry
use CHEMISTRY

Bioengineering
use BIOLOGICAL SCIENCES
ENGINEERING

BIOGRAPHICAL DOCUMENTS
X Autobiographies
Curricula Vitae
Obituaries
Resumes (Of Persons)
BT Documents

Biological Abstracts (Publication)
use ABSTRACTS PUBLICATIONS
BIOLOGICAL SCIENCES

BIOLOGICAL SCIENCES
BT Science And Technology
NT Agriculture
Biomedical Sciences

Ecology
Genetics
RT Chemistry
Cybernetics
Physical Sciences
PX AIBS (American Institute Of Biological Sciences)
American Institute Of Biological Sciences (AIBS)
Bioengineering
Biological Abstracts (Publication)
Biologists
Biosciences Information Service (BIOSIS) (Tradename)
BIOSIS (Biosciences Information Service) (Tradename)
CBAC (Chemical-Biological Activities) (Index)
CBE (Conference Of Biological Editors)
Conference Of Biological Editors (CBE)

Biologists
use BIOLOGICAL SCIENCES
SCIENTISTS

Biomedical Communications Network (BCN)
use BIOMEDICAL SCIENCES
TIME-SHARING SERVICES

BIOMEDICAL SCIENCES
X Cancer
Medical Sciences
Nutrition
Pharmaceutical Sciences
Pharmacology
Public Health
BT Biological Sciences
NT Health Care
Neurology
RT Behavioral Sciences
PX Aim (Abridged Index Medicus)
AMA (American Medical Association)
American Medical Association (AMA)
BCN (Biomedical Communications Network)
Biomedical Communications Network (BCN)
Chemotherapy
Excerpta Medica
Index Medicus

Medical Data
Medical Education
Medical Journals
Medical Library Association (MLA)
MEDLARS Project (Medical Literature Analysis and Retrieval System)
MLA (Medical Library Association)
Physicians
Regional Medical Libraries

Biosciences Information Service (BIOSIS) (Tradename)
use BIOLOGICAL SCIENCES
INFORMATION SCIENCE SERVICE ORGANIZATIONS

BIOSIS (Biosciences Information Service) (Tradename)
use BIOLOGICAL SCIENCES
INFORMATION SCIENCE SERVICE ORGANIZATIONS

Bits
use CODING SYSTEMS

Blacks
use ETHNIC GROUPS

Blind Persons
use HANDICAPPED PERSONS

BLUEPRINTS
BT Graphic Records

BNB (British National Bibliography)
use BIBLIOGRAPHY (PROCESS)
UNITED KINGDOM

Boards (Of Governance)
use ADMINISTRATION
STRUCTURE

BODY LANGUAGE
X Posture (Arranging Of The Human Body)
BT Symbol Sets
NT Dancing
RT Kinesics
Linguistics
Signals
Visual Communication

Book Catalog
  use STRUCTURE
    INDEXES

Book Circulation
  use DELIVERY OF DOCUMENTS
    BOOKS

BOOKS
    Used For Non-serial Publications Not Individually Identified As Members Of "Documents" Within This Thesaurus.
  X Guides (Reference Books)
    Handbooks
    Manuals
    Texts (Books)
  BT Documents
  PX Back-Of-The-Book Indexes
    Book Circulation
    Recorded Books
    Reserve Books
    Style Manuals

BOOLEAN ALGEBRA
  X Logical Product
    Logical Sum
  BT Algebras
  RT Symbolic Logic

Bound Terms
  use INDEX TERMS (FORMAT AND INTERRELATIONS)

Bradford's Law
  use STATISTICS
    USES (OF RESOURCES)

BRAILLE
  BT Alphabets

British (People)
  use UNITED KINGDOM
    POPULATION

British National Bibliography (BNB)
  use BIBLIOGRAPHY (PROCESS)
    UNITED KINGDOM

Broadcasting

use DISSEMINATING

Brochures
use PAMPHLETS

BROWSABILITY
BT Interaction

BROWSING
BT Associating
RT Availability

BUDGETING
BT Finance
RT Accounting
Acquiring
PX PPBS (Program Planning And Budgeting System)
Program Planning And Budgeting System (PPBS)

Building
use CONSTRUCTING

BUILDINGS
BT Facilities
NT Hospitals
Prisons
Residential Facilities
RT Architecture
Constructing
Libraries
Museums
PX Library Standards (Building Standards)

Bulletins
use NEWS PUBLICATIONS

Business (Subject Area)
use ECONOMICS

Business Administration
use ECONOMICS
ADMINISTRATION

Business Information Systems
use MANAGEMENT INFORMATION SYSTEMS

Business Organizations
  use MANUFACTURING AND SALES ORGANIZATIONS

BYLAWS
  BT Laws
  RT Documents
     Law
     Organizations
     Specifications

Bytes
  use CODING SYSTEMS

C

Cable Television (CATV)
  use TELEVISION

Cablegrams
  use CORRESPONDENCE (DOCUMENTS)

CAI (Computer Assisted Instructing)
  use COMPUTER ASSISTED INSTRUCTING (CAI)

CALCULATIONS
  BT Numeric Processing
  RT Data Processing Services

CALCULATORS
  BT Computers

Call Number
  use INDEX TERMS (DOCUMENT SURROGATES)

Cameras
  use PHOTOGRAPHY
     EQUIPMENT

CANADA
  BT Geographic Areas

Cancer
  use BIOMEDICAL SCIENCES

Capability
  use CAPACITY

CAPACITY
  X   Capability
       Incapability
  BT   Measures (Measuring Techniques)
  RT   Productivity
  PX   Capacity, Storage
       Storage Medium Capacity

Capacity, Storage
  use CAPACITY
       STORAGE SPACE

Card Catalogs
  use STRUCTURE
       INDEXES

Card Reproduction
  use CARDS
       REPRODUCING (OF DOCUMENTS)

CARDS
  X   Catalog Cards
  BT   Storage Media
  NT   Punched Cards
  PX   Card Reproducton
       Cards, Subject
       Optical Coincidence Techniques
       Subject Cards

Cards, Subject
  use CARDS
       INDEXES

Careers
  use EMPLOYMENT

Cartridges
  use STORAGE MEDIA

Cassettes
  use STORAGE MEDIA

Catalog Cards
  use CARDS

Cataloging, Descriptive
  use INDEXING
       CITATIONS

Catalogs
  use INDEXES
      BIBLIOGRAPHIES

Catalogs, Union
  use COOPERATION
      INDEXES

Categories
  use CLASSIFICATION SYSTEMS

CATHODE RAY TUBES
  BT Television Equipment
  RT Components

CATV (Cable Television)
  use TELEVISION

Causes
  use INTERACTION

CBAC (Chemical-Biological Activities) (Index)
  use INDEXES
      CHEMISTRY
      BIOLOGICAL SCIENCES

CBE (Conference Of Biological Editors)
  use BIOLOGICAL SCIENCES
      INFORMATION SCIENCE PROFESSIONAL
          ORGANIZATIONS

CENSORSHIP
  X Classification (Restriction By Government
          Regulation)
      Clearance (Of Documents)
  BT Control Functions
  PX Classified Documents (Restricted By
          Government Regulation)

Census
  use DATA
      POPULATION

CENSUS BUREAU
  BT Department Of Commerce

Centers
  use ORGANIZATIONS

CENTRAL INTELLIGENCE AGENCY (CIA)
  X   CIA (Central Intelligence Agency)
  BT   United States Government

Centralized Information Services
  use INFORMATION SCIENCE SERVICE ORGANIZATIONS

Centralizing
  use INTERRELATIONS

Certification
  use AWARDS

CFSTI (Clearinghouse For Federal Scientific And Technical Information)
  use INFORMATION SCIENCE SERVICE ORGANIZATIONS
       UNITED STATES GOVERNMENT
       SCIENCE AND TECHNOLOGY

Chadless Tape
  use PAPER TAPE

Chain Indexing
  use COORDINATE INDEXING
       PERMUTATION

Chained Files
  use FILE ORGANIZATION

CHANGE
  X   Adapting
       Innovation
       Modify
       Remedy
       Transition
       Upgrade
  BT   Control Functions
  NT   Growth
  RT   Interrelations
  PX   Rehabilitation
       Reversal

Channel (Hardware)
  use INPUT-OUTPUT EQUIPMENT

Channels Of Communication
  use INFORMATION FLOW

Character Recognition
  use RECOGNITION
      SYMBOL SETS

Character Recognition Equipment
  use OPTICAL SCANNERS

Characteristic (Indexing)
  use INDEX TERMS (DOCUMENT SURROGATES)

Characters
  use SYMBOL SETS

Charges (Financial)
  use COSTS

Charts
  use GRAPHIC RECORDS

Checkout (Of Documents)
  use DELIVERY OF DOCUMENTS

Chemical Abstracts (Publication)
  use ABSTRACTS PUBLICATIONS
      CHEMISTRY
      INDEXES

CHEMICAL COMPOUNDS
  X Chemical Structures
    Compounds, Chemical
  BT Chemistry
  PX Chemical Compounds Registry System

Chemical Compounds Registry System
  use CHEMICAL COMPOUNDS
      CODING SYSTEMS

Chemical Data
  use CHEMISTRY
      DATA

Chemical Documentation
  use CHEMISTRY
      DOCUMENTS
      STORAGE AND RETRIEVAL PROCESSES

Chemical Engineering
  use CHEMISTRY

ENGINEERING

Chemical Indexing
use CHEMISTRY
INDEXING

Chemical Literature
use CHEMISTRY
DOCUMENTS

Chemical Patents
use CHEMISTRY
PATENTS

Chemical Structures
use CHEMICAL COMPOUNDS

CHEMISTRY
X Biochemistry
Chemistry, Organic
Organic Chemistry
BT Science And Technology
NT Chemical Compounds
RT Biological Sciences
Engineering
Physical Sciences
PX ACS (American Chemical Society)
American Chemical Society (ACS)
CBAC (Chemical-Biological Activities) (Index)
Chemical Abstracts (Publication)
Chemical Data
Chemical Documentation
Chemical Engineering
Chemical Indexing
Chemical Literature
Chemical Patents
Chemists, Literature
Data, Chemical
Engineering, Chemical
Literature Chemists

Chemistry, Organic
use CHEMISTRY

Chemists, Literature
use CHEMISTRY
INFORMATION SCIENCE PERSONNEL

Chemotherapy
  use BIOMEDICAL SCIENCES
      DRUGS

Chicanos
  use ETHNIC GROUPS

CHILDREN
  BT Roles (Social)
  RT Adults
      Young Adults

CHINA
  BT Asia

CHINESE LANGUAGE
  BT Natural Languages

Choice
  use DECISION-MAKING

Chronic Dysfunctions
  use DYSFUNCTIONS

CIA (Central Intelligence Agency)
  use CENTRAL INTELLIGENCE AGENCY (CIA)

CIM (Computer Input Microforms)
  use COMPUTER APPLICATIONS

CIRCUITS
  BT Components
  PX Logical Circuitry

Circulation (Of Documents)
  use DELIVERY OF DOCUMENTS

Citation (Process)
  use BIBLIOGRAPHY (PROCESS)

CITATION INDEXES
  X Bibliographic Coupling
      Coupling, Bibliographic
  BT Documents
  RT Bibliographies
      Citations
      Indexes

Citation Studies
  use BIBLIOGRAPHIES
      EVALUATION

Citation, Literature
  use BIBLIOGRAPHIES

CITATIONS
  X Bibliographic Data
    Bibliographic Elements
    Intercitations
    References (To Documents)
  BT Bibliographies
  RT Bibliography (Process)
     Citation Indexes
  PX Cataloging, Descriptive

Citations (Bibliographies)
  use BIBLIOGRAPHIES

Cities
  X Metropolitan
    Municipalities
    Urban
  BT United States Of America
  RT Geographic Areas
     Regional
     Urban Government
  PX City Planning

Citizens
  use POPULATION

City Planning
  use CITIES
      PLANNING

Civic Groups
  use COMMUNITIES
      SERVICE ORGANIZATIONS

Civil Liberties
  use RIGHTS

Civil Service
  use EMPLOYMENT
      UNITED STATES GOVERNMENT

Classes (Logical)
  use CLASSIFICATION SYSTEMS

Classes (Social)
  use SOCIOECONOMIC GROUPS

Classification (Process)
  use APPLICATIONS
    CLASSIFICATION SYSTEMS

Classification (Restriction By Government Regulation)
  use CENSORSHIP

Classification Numbers
  use CODING SYSTEMS

CLASSIFICATION SYSTEMS
  X Categories
    Classes (Logical)
    Hierarchy
    Indexing Languages
    Nomenclature
    Registry Systems
    Subject Classification
    Taxonomy
    Tree Structures
  BT Arrangement
  NT Faceted Classification Systems
    Library Classification Systems
  RT Arranging
    Interrelations
  PX Classification (Process)

Classificationists
  use INDEXING
    INFORMATION SCIENCE PERSONNEL

Classified Documents (Restricted By Government Regulation)
  use CENSORSHIP
    DOCUMENTS

Classified Index
  use INDEXES
    ARRANGEMENT

Classifier
  use INDEXING

INFORMATION SCIENCE PERSONNEL

Clearance (Of Documents)
  use CENSORSHIP

Clearinghouse (Information)
  use INFORMATION SCIENCE SERVICE ORGANIZATIONS

Clearinghouse For Federal Scientific And Technical Information (CFSTI)
  use INFORMATION SCIENCE SERVICE ORGANIZATIONS
  UNITED STATES GOVERNMENT
  SCIENCE AND TECHNOLOGY

CLIENTELE
  X Recipients (Receivers) (Human)
  BT Roles (Social)
  NT Users (Of Services)
  PX Aides
  Referral

Climate
  use CONDITIONS

Clinic (Workshop)
  use MEETINGS (COMMUNICATING VIA MEETINGS)

Closed-Circuit Television
  use TELEVISION

Clubs (Social)
  use GROUPS

Cluster Analysis
  use FACTOR ANALYSIS

COBOL (PROGRAMMING LANGUAGE)
  BT Artificial Languages

Coden
  use PERIODICALS
  CODING SYSTEMS

Codes
  use CODING SYSTEMS

CODING
  X Encoding

Numeric Coding
BT Translation
RT Coding Systems
PX Automated Coding
Information Coding
Random Coding

Coding Fields
use APPLICATIONS
CODING SYSTEMS

CODING SYSTEMS
X Abbreviations
Acronym
Alphabetic Codes
Alphanumeric Codes
Binary Codes
Bits
Bytes
Classification Numbers
Codes
Gray Codes
Notation Systems
BT Symbol Sets
RT Arranging
Coding
Interrelations
Linguistics
Mathematics
Semantics
PX American Standard Code For Information Interchange (ASCII)
ASCII (American Standard Code For Information Interchange)
Chemical Compounds Registry System
Coden
Coding Fields
Genetic Coding System
Superimposed Coding

COGNITION
X Discrimination (Intellectual)
Perception
BT Intellect (Intelligence)
NT Concepts
RT Learning
Memory

Coincidence
  use CORRELATING

Cold (Temperature)
  use TEMPERATURE

Collaboration
  use COOPERATION

COLLATING
  BT Interrelations
  RT Comparing
     Electromechanical Data Processing Equipment
     Merging
     Sorting
  PX Collator (Machine)

Collator (Machine)
  use COLLATING
     EQUIPMENT

Collecting
  use ACQUIRING

Colleges
  use ACADEMIC ORGANIZATIONS

Colon Classification System
  use FACETED CLASSIFICATION SYSTEMS

Color Television
  use TELEVISION

COM (Computer Output Microforms)
  use COMPUTER APPLICATIONS

Commissions (Administrative Groups)
  use ADMINISTRATION
     INTERRELATIONS

Commitment
  use ATTITUDES

Committee Z 39
  use INFORMATION SCIENCE SERVICE ORGANIZATIONS
     SPECIFICATIONS

Committees

use GROUPS

COMMUNICATION
X Communication Science
BT Processes (Processing)
NT Instructing
Oral Communication
Person-To-Person Communication
Publishing
Recording
Translation
Transmission
Visual Communication
RT Cybernetics
Data Processing
Delivery Of Services
Relations
PX Communication Behavior
Communication Dysfunctions
Discussion
FCC (Federal Communications Commission)
Federal Communications Commission (FCC)
Group Communication
Human Communication
Information Exchange (Interpersonal)
Interpersonal Communication
Many-To-One Communications
One-To-Many Communication
Transmitters (Human)

Communication Behavior
use COMMUNICATION
BEHAVIOR

Communication Dysfunctions
use DYSFUNCTIONS

Communication Equipment
use INPUT-OUTPUT EQUIPMENT

Communication Networks (Automated)
use TIME-SHARING SERVICES

COMMUNICATION SATELLITES
BT Equipment
RT Communication Techniques
Telecommunication

Communication Science
  use   COMMUNICATION

COMMUNICATION SERVICES
  BT   Delivery Of Services

COMMUNICATION TECHNIQUES
  BT   Techniques
  NT   Microwave Techniques
       Multimedia Techniques
       Programs
       Telecommunication
  RT   Communication Satellites

Communication Theory
  use   INFORMATION SCIENCE
        THEORY

COMMUNITIES
  X   Ghetto
       Neighborhood
       Settlements
  BT   United States Of America
  PX   Civic Groups

Companies
  use   MANUFACTURING AND SALES ORGANIZATIONS

COMPARING
  X   Matching
  BT   Interrelations
  RT   Collating
       Compatibility
       Correlating
       Criteria
       Optical Coincidence Techniques
       Searching
       Selection
       Sorting

COMPATIBILITY
  BT   Interaction
  RT   Comparing
       Convertibility

COMPENSATION
  X   Remunerate
       Rewards

Workmen's Compensation
BT Employment
NT Awards
Insurance (Life)
Retirement Plans
Social Security
Wages
RT Finance
Financial Aid
Motivation
Uses (Of Resources)

Compiler Languages
use ARTIFICIAL LANGUAGES

COMPILING
BT Writing
RT Constructing

Compiling (Bibliographies)
use BIBLIOGRAPHY (PROCESS)

Compiling (Computer Programs)
use PROGRAMS
TRANSLATION

Complexity
use CONDITIONS

COMPONENTS
X Switching Equipment
BT Equipment
NT Circuits
Lenses
RT Cathode Ray Tubes

COMPOSING
X Digital Plotting
Formatting
Typography
BT Interrelations
NT Photocomposition
RT Constructing
Printing
Printing Equipment
Publications (Personal Or Organizational)
Publishing
PX Computer Graphics

Format Standardization
Graphics (Composing)

Composition Equipment
use PRINTING EQUIPMENT

Compounds, Chemical
use CHEMICAL COMPOUNDS

Compression (Of Data)
use DATA
TRANSLATION

Computation
use NUMERIC PROCESSING

Computational Linguistics
use LINGUISTICS
COMPUTER APPLICATIONS

COMPUTER APPLICATIONS
Note: "Electronic Digital Computers" Is Abbreviated to "Computers" In This Term.
X Applications, Computer
Automation
BT Applications
RT Linguistics
Time-Sharing Services
Word Processing
PX Autoabstracting
Automated Coding
Automated Indexing
Automated Language Processing
Automated Language Translation
Automated Law Enforcement
Automated Medical Diagnosis
Automated Medical History Systems
Automated Retrieval
Automated Subject Analysis
Automated Typesetting
CIM (Computer Input Microforms)
COM (Computer Output Microforms)
Computational Linguistics
Computer Assisted Medical Record Systems
Computer Graphics
Computer Input Microforms (CIM)
Computer Modeling

Computer Output Microforms (COM)
Computer Scheduling
Computer Simulation
Language Processing (Computers)
Language Translation (Automated)
Machine Translation
Mechanized Literature Searching
Question-Answering Systems (Automated)

Computer Assisted Indexing
use INDEXING
MAN-MACHINE INTERACTION

COMPUTER ASSISTED INSTRUCTING (CAI)
X CAI (Computer Assisted Instructing)
Electronic Learning Packages
BT Instructing
NT Plato System

Computer Assisted Literature Alerting
use INFORMATION FLOW
TECHNIQUES

Computer Assisted Medical Diagnosis
use MAN-MACHINE INTERACTION
HEALTH CARE

Computer Assisted Medical Record Systems
use MEDICAL RECORDS
COMPUTER APPLICATIONS

Computer Graphics
use GRAPHIC RECORDS
COMPOSING
COMPUTER APPLICATIONS

Computer Input Devices
use INPUT-OUTPUT EQUIPMENT

Computer Input Microforms (CIM)
use COMPUTER APPLICATIONS
MICROFORM DOCUMENTS

Computer Memory
use MEMORY

Computer Modeling
use MODELING

COMPUTER APPLICATIONS

Computer Networks
use TIME-SHARING SERVICES

Computer Output Devices
use INPUT-OUTPUT EQUIPMENT

Computer Output Microforms (COM)
use COMPUTER APPLICATIONS
MICROFORM DOCUMENTS

Computer Programming
use PROGRAMMING

Computer Scheduling
use SCHEDULING
COMPUTER APPLICATIONS

Computer Simulation
use MODELING
COMPUTER APPLICATIONS

Computer Software
use PROGRAMS

Computer Storage
use STORAGE MEDIA

Computer Tape
use MAGNETIC TAPE

COMPUTER TERMINALS
X Terminals (Computer)
BT Input-Output Equipment

Computer Utilities
use TIME-SHARING SERVICES

COMPUTERS
BT Data Processing Equipment
NT Analog Equipment
Calculators
Electromechanical Data Processing Equipment
Electronic Digital Computers
Hybrid Computers
RT Applications
Electronics

Concept Coordination
  use CONCEPTS
      COORDINATION

CONCEPTS
  X Content
     Ideas
     Message (Content Of A Communication)
  BT Cognition
  NT Theory
  RT Coordinate Indexing
     Index Terms (Document Surrogates)
     Learning
     Logic
     Semantics
  PX Concept Coordination
     Content Analysis
     Content Recognition
     Document Analysis
     Document Clustering Analysis
     Question-Answering Systems (Automated)

Concordances
  use INDEXES

Condensations (Of Literature)
  use ABSTRACTS (OF DOCUMENTS)

CONDITIONING
  X Reinforcement
  BT Behavior

CONDITIONS
  X Climate
     Complexity
     Dynamics
     Living Conditions
     Milieu
     Parameters
     Situation
  BT Environment
  NT Crisis
     Floods
  RT Processes (Processing)
     Resources

Conference Calls (Telephone)
  use ORAL COMMUNICATION

TELEPHONE

Conference Of Biological Editors (CBE)
use BIOLOGICAL SCIENCES
INFORMATION SCIENCE PROFESSIONAL ORGANIZATIONS

Conferences
use MEETINGS (COMMUNICATING VIA MEETINGS)

Congress (U.S.)
use UNITED STATES CONGRESS

Congresses
use MEETINGS (COMMUNICATING VIA MEETINGS)

Conjunctions (Logical)
use LOGIC

CONSISTENCY
BT Interrelations
PX Indexer Consistency

Consorting (Consortia)
use COOPERATION

CONSTRUCTING
X Assembling
Building
BT Interrelations
RT Buildings
Compiling
Composing
Designing
PX Report Generators

Consultants
use CONSULTING SERVICES
PERSONNEL

CONSULTING SERVICES
BT Delivery Of Services
NT Counseling
RT Academic Organizations
Human Relations
Personnel
Professional Organizations
Service Organizations

Users (Of Resources)
PX Consultants
Experts

Consumer Education
use CONSUMERS
INSTRUCTING

CONSUMERS
BT Roles (Social)
NT Users (Of Services)
RT Advertising
PX Consumer Education

Content
use CONCEPTS

Content Analysis
use LINGUISTICS
CONCEPTS

Content Recognition
use RECOGNITION
CONCEPTS

Content-Addressable Memories
use ASSOCIATING
MEMORY

Contents Pages (Tables Of Contents Of Documents)
use INDEXES

Continuing Education
use POSTGRADUATE EDUCATION

Continuous Records
use STORAGE MEDIA
ARRANGEMENT

Contract Proposals
use GRANTS AND CONTRACTS
PROPOSALS

Contracts
use GRANTS AND CONTRACTS

CONTROL FUNCTIONS
X Authority

Controlling
Discarding
Discharging (Persons)
Facilitating
Licensing
Ownership
Policing
Validation
BT Administration
NT Censorship
Change
Decision-Making
Identification
RT Acquiring
Bibliography (Process)
Governments
Monitoring
Operations Research
Responsibility
Rules
Space (Resource)
PX Data Control
Elective Offices
Firing
Hiring
Quality Control
Quantity Control
Refereeing

Controlled Vocabularies
use THESAURI

Controlling
use CONTROL FUNCTIONS

Controls (Research)
use POPULATION

Conventional Systems (Of Indexing)
use INDEXES
ARRANGEMENT

Conversion
use TRANSLATION

Converter
use TRANSLATION
EQUIPMENT

CONVERTIBILITY
X Data Interchange Standardization
BT Interaction
RT Compatibility
Exchanging
Translation
PX Data Convertibility

COOPERATION
X Collaboration
Consorting (Consortia)
BT Interaction
NT Exchanging
RT Compatibility
Coordination
PX Catalogs, Union
Cooperative Library Programs
Union Catalogs
Union Lists

Cooperative Library Programs
use LIBRARIES
COOPERATION

COORDINATE INDEXING
X Correlative Indexing
Faceted Indexing
BT Indexing
RT Concepts
Correlating
Faceted Classification Systems
PX Chain Indexing

COORDINATION
BT Interaction
RT Cooperation
PX Concept Coordination

Coordinators
use ADMINISTRATION
PERSONNEL

Copies (Of Documents)
use REPRODUCTIONS

Copying (Of Documents)
use REPRODUCING (OF DOCUMENTS)

COPYING SERVICES
  X   Document Reproduction Services
  BT   Delivery Of Services
  RT   Printing Services

COPYRIGHT LAW
  BT   Laws
  PX   Fair Use Copying

Core Memory
  use MEMORY

Corporations
  use MANUFACTURING AND SALES ORGANIZATIONS

CORRELATING
  X   Autocorrelation
       Coincidence
  BT   Statistics
  RT   Associating
       Comparing
       Coordinate Indexing

Correlative Indexing
  use COORDINATE INDEXING

CORRESPONDENCE (DOCUMENTS)
  X   Cablegrams
       Letters (Correspondence)
       Mailgrams
       Telegrams
  BT   Documents
  RT   Records
  PX   Radiograms

Cost Accounting
  use COSTS
       ACCOUNTING

Cost Analysis
  use COSTS
       ACCOUNTING

Cost Benefit
  use COSTS
       BENEFIT

Cost Effectiveness

use COSTS
EFFECTIVENESS

COSTS
X Charges (Financial)
Fees
Inexpensive
Purchase Cost
BT Finance
RT Acquiring
Designing
Effectiveness
Efficiency
Feasibility
Measures (Measuring Techniques)
PX Cost Accounting
Cost Analysis
Cost Benefit
Cost Effectiveness
Price/Value

COUNSELING
BT Consulting Services

Countries
use GEOGRAPHIC AREAS

Coupling, Bibliographic
use CITATION INDEXES

COURSE WRITER (PROGRAMMING LANGUAGE)
BT Artificial Languages

COURSES (EDUCATIONAL)
X Lessons
BT Instructing
RT Curricula
High School Education
Postgraduate Education

COURTS
BT Laws

Cranfield Project
use INDEXING
EVALUATION

Creative Writing

use WRITING (CREATIVE)

CREATIVITY
X Imagination
Synthesize
BT Ability
NT Designing
Modeling
Writing (Creative)
RT Planning

CRISIS
X Emergency
BT Conditions
RT Dysfunctions

Crisis, Information
use INFORMATION FLOW
DYSFUNCTIONS

CRITERIA
X Standards
BT Values
NT Indicators
Specifications
RT Designing
Evaluation
Recognition
Reviews
Selection
PX Critical Reviews
Criticism
Data Interchange Standardization
Federal Information Processing Standards
Format Standardization
International Standards Organization (ISO)
ISO (International Standards Organization)
Library Standards (Building Standards)
Library Standards (For Services)
References (For Employment)
Standardization

CRITICAL INCIDENT METHOD
BT Data Gathering

Critical Reviews
use CRITERIA
REVIEWS

Criticism
  use CRITERIA
      APPLICATIONS

Cross Referencing
  use INDEX TERMS (FORMAT AND INTERRELATIONS)

Crystallography
  use PHYSICAL SCIENCES

CULTURES
  X Society
  BT Groups
  RT Anthropology
      Socioeconomic Groups

Cumulated Indexes
  use INDEXES
      MERGING

Cumulation
  use MERGING

Currency (Of Information)
  use TIME

Current Awareness
  use INFORMATION FLOW
      TIME

CURRICULA
      Note: Used For Courses Of Study Leading To An Academic Degree.
  BT Instructing
  RT Courses (Educational)
      High School Education
      Postgraduate Education
      Programs

Curricula Vitae
  use BIOGRAPHICAL DOCUMENTS

CYBERNETICS
  X Systems Theory
  BT Physical Sciences
  RT Biological Sciences
      Communication
      Electronics

Engineering
Human Beings
Mathematics

CZECHOSLOVAKIA
BT Europe

D

DANCING
BT Body Language
RT Arts And Humanities
Music (Art Form)
Singing
Visual Communication

DATA
Note: Used Synonymously With "Information" In Many Instances.
X Files
Input Data
Laboratory Data
BT Information
NT Demographic Data
Machine Readable Data
RT Data Processing
Documents
Electromechanical Data Processing Equipment
Input Records
Interviewing
Knowledge (State Of Being Informed)
Libraries
Research Results
PX Analog-Digital Conversion
Census
Chemical Data
Compression (Of Data)
Data Compression
Data Control
Data Conversion
Data Convertibility
Data Description Languages
Data Interchange Standardization
Data Mapping
Data Transmission
Data, Chemical

File Compression
File Design
File Maintenance
File Security
Graphic Data
Graphic Information
Information Coding
Information Collection (Noun) (Pool Of Data)
Information Exchange (Data)
Medical Data

DATA ANALYSIS
X Analysis, Data
Analysis, Information
Analysis, Statistical
BT Evaluation Techniques
NT Factor Analysis
PERT (Program Evaluation Review Techniques)
RT Data Gathering
Interrelations
Research Results
Statistics

Data Banks
use STORAGE AND RETRIEVAL SYSTEMS

Data Bases (For Automated Storage And Retrieval)
use INDEXES
STORAGE AND RETRIEVAL PROCESSES
SEARCHING

Data Compression
use DATA
TRANSLATION

Data Control
use CONTROL FUNCTIONS
DATA

Data Conversion
use DATA
TRANSLATION

Data Convertibility
use DATA
CONVERTIBILITY

Data Description Languages
  use INTERRELATIONS
      ARTIFICIAL LANGUAGES
      DATA

DATA GATHERING
  X Information Collecting (Data)
     Information Gathering
     Retrospective Searching
  BT Evaluation Techniques
  NT Critical Incident Method
     Delphi Method
     Diary Techniques
     Questionnaire Techniques
     Random Alarm Techniques
     Selection
     Surveying
  RT Acquiring
     Data Analysis
     Sampling

Data Interchange Standardization
  use DATA

Data Mapping
  use DATA
      TRANSLATION

Data Phone (Tradename)
  use DATA PROCESSING
      TELEPHONE

DATA PROCESSING
     Note: Used For Both Human And Automated
           Processing Of Data Or Information.
  X Automated Information Processing
     EDP (Electronic Data Processing)
     Electronic Data Processing (EDP)
     Information Processing
     Mechanized Information Processing
  BT Processes (Processing)
  NT Batch Processing
     Error Detection And Correction
     Numeric Processing
     Offline Data Processing
     Online Data Processing
     Storage And Retrieval Processes
     Word Processing

RT Communication
Data
Delivery Of Services
Electronic Digital Computers
Information
Input Records
Uses Of Resources
PX ACM (Association For Computing Machinery)
Afips (American Federation Of Information Processing Societies)
American Federation Of Information Processing Societies (AFIPS)
American Standard Code For Information Interchange (ASCII)
ASCII (American Standard Code For Information Interchange)
Association For Computer Machinery (ACM)
Data Phone (Tradename)
Federal Information Processing Standards
International Federation Of Information Processing (IFIP)
Language Processing (Computers)
List Processing
Multiprocessing
Real-Time

Data Processing Centers
use DATA PROCESSING SERVICES

DATA PROCESSING EQUIPMENT
X EDP Equipment
Information Handling Equipment
Word Processing Equipment
BT Equipment
NT Computers
Input-Output Equipment
RT Electronics
PX Data Processing Systems

DATA PROCESSING SERVICES
X Data Processing Centers
BT Delivery Of Services
NT Word Processing Services
RT Bibliographies
Graphic Records
Numeric Processing
Word Processing

Data Processing Systems
  use APPLICATIONS
  DATA PROCESSING EQUIPMENT

DATA PROCESSING TECHNIQUES
  X Library Science Techniques
  RT Techniques
  NT Optical Coincidence Techniques
  Programming
  Time-Sharing

Data Transmission
  use TRANSMITTERS (EQUIPMENT)
  DATA

Data, Chemical
  use CHEMISTRY
  DATA

Data, Graphic
  use MEDICAL RECORDS

Data, Input
  use INPUT RECORDS

Data, Medical
  use MEDICAL RECORDS

DAY
  BT Time

DDC (Defense Documentation Center)
  use DEFENSE DOCUMENTATION CENTER (DDC)

DDD (Direct Distance Dialing)
  use ACCESSING

Deaf Persons
  use HANDICAPPED PERSONS

Debts
  use FINANCE

Decentralization
  use INTERRELATIONS

Decision Theory
  use DECISION-MAKING

THEORY

DECISION-MAKING
X Choice
Judging
BT Control Functions
NT Planning
RT Designing
Evaluation
Input Records
Resources
PX Decision Theory

Defects
use DYSFUNCTIONS

DEFENSE DOCUMENTATION CENTER (DDC)
X DDC (Defense Documentation Center)
BT Department Of Defense
RT Libraries
Military Science

DEFENSE INTELLIGENCE AGENCY
BT Department Of Defense

Deficiencies
use DYSFUNCTIONS

Definition
use IDENTIFICATION

Degrees (Academic)
use AWARDS
ACADEMIC ORGANIZATIONS

Delivery (Transport Of Goods)
use TRANSPORTATION

DELIVERY OF DOCUMENTS
X Checkout (Of Documents)
Circulation (Of Documents)
Document Delivery
Document Dissemination
Loan (Of Library Materials)
BT Delivery Of Services
PX Book Circulation

DELIVERY OF SERVICES
X Library Services
BT Responses
NT Audiovisual Production Services
Binding Services
Communication Services
Consulting Services
Copying Services
Data Processing Services
Delivery Of Documents
Editing Services
Educational Services
Equipment Provision
Language Translation Services
Printing Services
Publishing Services
Responses (To Inquiries)
Space Provision
RT Communication
Data Processing
Library Science
Service Organizations
Techniques
Transmission

DELPHI METHOD
BT Data Gathering
RT Statistics

DEMOGRAPHIC DATA
BT Data

DEMONSTRATION
BT Experimental Research

DENMARK
BT Europe

DENTISTRY
BT Health Care

DEPARTMENT OF COMMERCE
BT United States Government
NT Census Bureau
United States Patent Office

DEPARTMENT OF DEFENSE
BT United States Government

NT Defense Documentation Center (DDC)
Defense Intelligence Agency
United States Air Force
United States Army
United States Navy
RT Military Science

Dependent
use RELATIONS

Dependent Variables
use VARIABLES

Depreciation
use FINANCE

DEPRESSED AREAS
BT United States Of America
RT Geographic Areas
Socioeconomic Groups

Descriptive Cataloging (Process)
use INDEXING
BIBLIOGRAPHY (PROCESS)

Descriptors
use INDEX TERMS (DOCUMENT SURROGATES)

Design (Noun)
use SPECIFICATIONS

Designers
use DESIGNING
PERSONNEL

DESIGNING
BT Creativity
RT Applications
Benefit
Constructing
Costs
Criteria
Decision-Making
Effectiveness
Efficiency
Equipment
Evaluation
Feasibility

Specifications
PX Designers
System Design

DETERIORATION
BT Dysfunctions

Determinants
use INTERACTION

Development (Physical)
use GROWTH

Device (Plan)
use PLANNING

Devices
use EQUIPMENT

DEWEY DECIMAL CLASSIFICATION SYSTEM
BT Library Classification Systems

DIAGNOSIS
BT Evaluation
PX Automated Medical Diagnosis

Dial-Access Information Retrieval
use TELEPHONE

DIARY TECHNIQUES
BT Data Gathering
RT Surveying

DICTIONARIES
X Glossaries
Lexicon (Natural Language)
Vocabularies (Natural Language)
BT Documents
RT Thesauri

Dictionary Catalog
use INDEX TERMS (FORMAT AND INTERRELATIONS)

Digital Computers
use ELECTRONIC DIGITAL COMPUTERS

Digital Plotting
use COMPOSING

Digital Storage
use STORAGE MEDIA

Digital-Analog Computers
use HYBRID COMPUTERS

Direct Distance Dialing (DDD)
use ACCESSING
TECHNIQUES
TELEPHONE

Direct-Access Computing Techniques
use TIME-SHARING SERVICES

Directed Graphs
use STRUCTURE

Directing
use ADMINISTRATION

Directories
use INDEXES

Disabled Persons
use HANDICAPPED PERSONS

DISADVANTAGED PERSONS
X Socially Handicapped
BT Groups
RT Dysfunctions
Handicapped Persons

Disasters
use DYSFUNCTIONS

Discarding
use CONTROL FUNCTIONS

Discharging (Persons)
use CONTROL FUNCTIONS

Disciplines
use KNOWLEDGE AREAS

Discontinuous Records
use STORAGE MEDIA
ARRANGEMENT

Discrimination (Electronic Object Detection)
  use ELECTRONICS
      RECOGNITION

Discrimination (Intellectual)
  use COGNITION

Discrimination (Social)
  use ETHNIC GROUPS

Discs (Disks)
  use MAGNETIC DISCS

Discussion
  use GROUPS
      COMMUNICATION

Display Devices (Data Processing)
  use INPUT-OUTPUT EQUIPMENT

Displays (Noun)
  use EXHIBITING

DISPOSAL
  X Retirement (Of Goods)
  BT Uses (Of Resources)
  PX Weeding (Of Document Collections)

DISSEMINATING
  X Announcement Services
    Broadcasting
    Distributing (Data Or Documents)
    Information Dissemination
  BT Transmission
  NT Information Flow
    Routing
    SDI
  RT Designers
    News Publications
    Publishing
    Specifications
    Transmitters (Equipment)

DISSERTATIONS
  X Theses
  BT Documents

Distributing (Data Or Documents)

use DISSEMINATING

Document Analysis
use EVALUATION
CONCEPTS
DOCUMENTS

Document Clustering Analysis
use FACTOR ANALYSIS
CONCEPTS
DOCUMENTS

Document Delivery
use DELIVERY OF DOCUMENTS

Document Description
use INDEXING

Document Dissemination
use DELIVERY OF DOCUMENTS

Document Organization (Process)
use DOCUMENTS
INTERRELATIONS

Document Representation
use INDEXING

Document Reproduction Services
use COPYING SERVICES

Document Retrieval
use STORAGE AND RETRIEVAL PROCESSES

Document Storage
use DOCUMENTS
STORAGE AND RETRIEVAL SYSTEMS

Document Surrogates
use INDEX TERMS (DOCUMENT SURROGATES)

Document Surrogation
use INDEXING

Documentalists
use INFORMATION SCIENCE PERSONNEL

DOCUMENTARIES

BT Documents

Documentation (In Evidence)
use RECORDS

Documentation (Processing Of Documents)
use STORAGE AND RETRIEVAL PROCESSES
DOCUMENTS

Documentation Centers
use INFORMATION SCIENCE SERVICE ORGANIZATIONS

Documentation Literature
use INFORMATION SCIENCE
DOCUMENTS

DOCUMENTS
X Full Text
BT Information
NT Abstracts (Of Documents)
Advertising Literature
Audiovisual Software (Documents)
Bibliographies
Biographical Documents
Books
Correspondence (Documents)
Dictionaries
Dissertations
Documentaries
Exhibits (Documents)
Forecasts (Documents)
Identification (Documents Of)
Indexes
Lay Publications
Meeting Documents
Name And Address Lists
News Publications
Pamphlets
Patents
Periodicals
Proposals
Publications (Personal Or Organizational)
Records
Reproductions
Reviews
Speeches
Thesauri
RT Advertising
Bylaws

Data
Graphic Information
Input Records
Medical Records
Questionnaires
Storage Media
PX Chemical Documentation
Chemical Literature
Classified Documents (Restricted By Government Regulation)
Document Analysis
Document Clustering Analysis
Document Organization (Process)
Document Storage
Documentation (Processing of Documents)
Documentation Literature
Editorials
Grouping (Of Documents)
Historical Documents
Humorous Documents
Information Collecting (Documents)
Information Collection (Noun) (Group Of Documents)
Information Exchange (Documents)
Oral Histories
Professional Literature
Russian Literature
Tutorial Documents
Weeding (Of Documents)

Down-time
use DYSFUNCTIONS

Drama
use ARTS AND HUMANITIES

Drawings
use GRAPHIC RECORDS

Drug Information Systems
use STORAGE AND RETRIEVAL SYSTEMS
DRUGS

DRUGS
BT Substances
PX Chemotherapy
Drug Information Systems

Drums (Memory)
  use STORAGE MEDIA

Duplicating (Documents)
  use REPRODUCING (OF DOCUMENTS)

Duration
  use TIME

Dynamics
  use CONDITIONS

DYSFUNCTIONS
  X Chronic Dysfunctions
    Defects
    Deficiencies
    Disasters
    Down-Time
    Responses (Dysfunctional)
    Riots
  BT Output
  NT Deterioration
    Failure
    Noise (In Communication Systems)
  RT Crisis
    Disadvantaged Persons
    Handicapped Persons
    Maintenance
  PX Communication Dysfunctions
    Crisis, Information
    False Drops (Unwanted Selections)
    Information Crisis
    Language Problems
    Learning Problems
    Rehabilitation
    Vandalism

E

ECOLOGY
  BT Biological Sciences
  PX Air Pollution Information Center (APTIC)
    APTIC (Air Pollution Information Center)

ECONOMICS
  X Business (Subject Area)

Insurance (Subject Area)
BT Behavioral Sciences
NT Banking
RT Employment
Finance
Insurance (Life)
PX Business Administration
EEC (European Economic Community)

EDGE-NOTCHED CARDS
X Mcbee Cards
Zator Cards
BT Punched Cards

Editing
use ERROR DETECTION AND CORRECTION

EDITING SERVICES
BT Delivery Of Services
RT Error Detection And Correction

Editorials
use ATTITUDES
DOCUMENTS

Editors
use ERROR DETECTION AND CORRECTION
INFORMATION SCIENCE PERSONNEL

EDP (Electronic Data Processing)
use DATA PROCESSING

EDP Equipment
use DATA PROCESSING EQUIPMENT

EDUCATION
BT Behavioral Sciences
NT Elementary School Education
High School Education
Postgraduate Education
University And College Education
RT Academic Organizations
Instructing
Learning
PX Educational Resources Information Center (ERIC)
ERIC (Educational Resources Information Center)

Educational Institutions
  use ACADEMIC ORGANIZATIONS

Educational Resources Information Center (ERIC)
  use EDUCATION
      INFORMATION SCIENCE SERVICE ORGANIZATIONS

EDUCATIONAL SERVICES
  BT Delivery Of Services

Educational Television (ETV)
  use INSTRUCTING
      TELEVISION

Educators
  use INSTRUCTING
      INFORMATION SCIENCE PERSONNEL

EDUCOM (Educational Communications)
  use TIME-SHARING SERVICES
      ACADEMIC ORGANIZATIONS

EEC (European Economic Community)
  use ECONOMICS
      EUROPE

EFFECTIVENESS
  X Adequacy
    Utility
  BT Measures (Measuring Techniques)
  RT Benefit
    Costs
    Designing
    Efficiency
    Equipment
    Feasibility
    Interaction
    Reliability
    Work (Energy Output)
  PX Cost Effectiveness
    Price/Value
    Retrieval Effectiveness

Effects
  use INTERACTION

EFFICIENCY
  X Efficiency Ratings

BT Measures (Measuring Techniques)
RT Benefit
Costs
Designing
Effectiveness
Equipment
Feasibility
Interaction
Reliability
Work (Energy Output)

Efficiency Ratings
use EFFICIENCY

Election (To Office)
use AWARDS

Elective Offices
use ORGANIZATIONS
CONTROL FUNCTIONS

ELECTROMECHANICAL DATA PROCESSING EQUIPMENT
BT Computers
RT Accounting
Collating
Data
Electronic Digital Computers
Input-Output Equipment
Interrelations
Merging
Optical Coincidence Techniques
Punched Cards
Sorting

Electronic Data Processing (EDP)
use DATA PROCESSING

ELECTRONIC DIGITAL COMPUTERS
Note: Add To This Thesaurus The Names Of Individual Computers As Needed.
X Digital Computers
BT Computers
NT Microprocessors
Mini-Computers
RT Accounting
Analog Equipment
Automata
Data Processing

Electromechanical Data Processing Equipment
Hybrid Computers
Information Science Service Organizations
PX Automated Translation (Transformation)

Electronic Learning Packages
use COMPUTER ASSISTED INSTRUCTING (CAI)

Electronic Videorecording (EVR)
use VIDEOTAPES (DOCUMENTS)
RECORDING

ELECTRONICS
BT Physical Sciences
RT Computers
Cybernetics
Data Processing Equipment
Engineering
PX Discrimination (Electronic Object Detection)

ELECTROSTATIC PROCESSES
X Xerography (Tradename)
BT Reproducing (Of Documents)
RT Photocopying

ELEMENTARY SCHOOL EDUCATION
BT Education

Eligibility
use RULES

Emergency
use CRISIS

Emotions
use PERSONALITY

Employees
use PERSONNEL

EMPLOYMENT
X Careers
Jobs
Occupations
Professions
Recruiting
Work

BT Administration
NT Compensation
RT Economics
Personnel
Retirement
Skills
Work (Energy Output)
PX Civil Service
Firing
Hiring
Job Analysis
Job Description
Job Performance
Jobs
Labor Unions
Organized Labor
References (For Employment)
Sex (Differences In Employment Or Compensation)
Work Study Program

Encoding
use CODING

ENGINEERING
BT Physical Sciences
RT Chemistry
Cybernetics
Electronics
Sound
PX Bioengineering
Chemical Engineering
Engineering Drawings
Engineering, Chemical
Engineering, Human Factor
Engineers
Human Engineering
IEEE (Institute Of Electrical And Electronic Engineers)
Institute Of Electrical And Electronic Engineers (IEEE)

Engineering Drawings
use ENGINEERING
GRAPHIC RECORDS

Engineering, Chemical
use CHEMISTRY
ENGINEERING

Engineering, Human Factor
  use ENGINEERING
      HUMAN BEINGS

Engineers
  use ENGINEERING
      SCIENTISTS

ENGLISH LANGUAGE
  BT Natural Languages

Entry (Indexing)
  use INDEX TERMS (DOCUMENT SURROGATES)

Entry (To Gain Access)
  use ACCESSING

ENVIRONMENT
  X Settings
  BT Information Science
  NT Conditions
      Goals
      Relations
      Resources
      Roles (Social)
      Rules

ENVIRONMENTAL PROTECTION AGENCY (EPA)
  X EPA (Environmental Protection Agency)
  BT United States Government

EPA (Environmental Protection Agency)
  use ENVIRONMENTAL PROTECTION AGENCY (EPA)

EQUIPMENT
  BT Input
  NT Automata
      Communication Satellites
      Components
      Data Processing Equipment
      Instruments
      Lasers
      Motor Vehicles
      Storage Media
  RT Applications
      Designing
      Effectiveness
      Efficiency

Sound
Transmission

EQUIPMENT PROVISION
BT Delivery Of Services

ERIC (Educational Resources Information Center)
use EDUCATION
INFORMATION SCIENCE SERVICE ORGANIZATIONS

ERROR DETECTION AND CORRECTION
X Editing
Proofreading
Revision
Update
Verification
BT Data Processing
RT Editing Services
PX Editors
Readers (Proof-Readers)

ETHICS
BT Values
RT Human Beings
Rules
PX Professional Behavior

ETHNIC GROUPS
X Blacks
Chicanos
Discrimination (Social)
Mexican Americans
Race
Spanish Americans
BT Groups
RT Socioeconomic Groups

ETV (Educational Television)
use INSTRUCTING
TELEVISION

EUROPE
BT Geographic Areas
NT Czechoslovakia
Denmark
France
Germany
Hungary

Ireland
Italy
Norway
Poland
Russia
Spain
Sweden
Switzerland
Turkey
United Kingdom
PX EEC (European Economic Community)

EVALUATION
X Analyzing
Investigating
Measuring
Relevance
Reviewing Process
Testing
BT Feedback And Control
NT Diagnosis
Measures (Measuring Techniques)
Research
RT Criteria
Decision-Making
Designing
Inquiries
Reviews
Storage And Retrieval Processes
PX Citation Studies
Cranfield Project
Document Analysis
Investigators
Relevance Studies
State-Of-The-Art Studies
System Analysis
Trends

EVALUATION TECHNIQUES
BT Techniques
NT Data Analysis
Data Gathering
RT Abstracting
Indexing
Linguistics
Mathematics
Operations Research
Statistics

Evidence (Documents In Evidence)
  use RECORDS

Evolution
  use HISTORY

EVR (Electronic Videorecording)
  use VIDEOTAPES (DOCUMENTS)
      RECORDING

Excerpta Medica
  use INDEXES
      ABSTRACTS PUBLICATIONS
      BIOMEDICAL SCIENCES

EXCHANGING
  BT Cooperation
  NT Interlibrary Loan
  RT Compatibility
     Convertibility
  PX Information Exchange (Data)
     Information Exchange (Documents)
     Library Networks

Executives
  use ADMINISTRATION
      PERSONNEL

EXHIBITING
  X Displays (Noun)
    Samples (Display Objects; Tokens)
  BT Marketing
  RT Advertising
     Advertising Literature

EXHIBITS (DOCUMENTS)
  BT Documents
  RT Samples

EXPECTATIONS
  BT Goals
  RT Attitudes
     Values

Expenditures
  use FINANCE

EXPERIMENTAL RESEARCH

BT Research
NT Demonstration

Experts
use CONSULTING SERVICES
PERSONNEL

Exploitation
use USES (OF RESOURCES)

Explore
use RESEARCH

EXTRACTING
Note: Used In The Sense Of Selecting Verbatim Portions Of Text.
X Abridgement (Abridging)
BT Interrelations
RT Abstracting

Extracts
use ABSTRACTS (OF DOCUMENTS)

F

Facet Analysis
use FACTOR ANALYSIS

FACETED CLASSIFICATION SYSTEMS
X Colon Classification System
BT Classification Systems
RT Coordinate Indexing
Indexing

Faceted Indexing
use COORDINATE INDEXING

Facilitating
use CONTROL FUNCTIONS

FACILITIES
X Laboratories
Living Facilities
Parking Facilities
BT Resources
NT Buildings

Facsimile Transmission
  use TRANSMITTERS (EQUIPMENT)
      REPRODUCTIONS

Facsimiles
  use REPRODUCTIONS

FACTOR ANALYSIS
  X Cluster Analysis
    Facet Analysis
    Semantic Factoring
  BT Data Analysis
  PX Document Clustering Analysis

Faculty
  use INSTRUCTING
      PERSONNEL

FAILURE
  BT Dysfunctions

Fair Use Copying
  use COPYRIGHT LAW
      PHOTOCOPYING

False Drops (Unwanted Selections)
  use SELECTION
      DYSFUNCTIONS

FAMILY
  X Friends
  BT Roles (Social)

FBI (Federal Bureau Of Investigation)
  use LAW
      UNITED STATES GOVERNMENT

FCC (Federal Communications Commission)
  use COMMUNICATION
      LAW

FDA (United States Food And Drug Administration)
  use PUBLIC HEALTH SERVICE

FEASIBILITY
  BT Measures (Measuring Techniques)
  RT Benefit
     Costs

Designing
Effectiveness
Efficiency
Planning
Reliability

Federal Bureau Of Investigation (FBI)
use LAW
UNITED STATES GOVERNMENT

Federal Communications Commission (FCC)
use COMMUNICATION
LAW

Federal Government
use UNITED STATES GOVERNMENT

Federal Information Processing Standards
use DATA PROCESSING
CRITERIA
UNITED STATES GOVERNMENT

Federation Internationale de Documentation (FID)
use INTERNATIONAL (GEOGRAPHY)
INFORMATION SCIENCE SERVICE ORGANIZATIONS

FEEDBACK AND CONTROL
Note: Used Only When More Specific Terms Will Not Suffice.
X Problem-solving
BT Information Science
NT Administration
Evaluation
RT Inquiries

Fees
use COSTS

Fellowships
use FINANCIAL AID
POSTGRADUATE EDUCATION

Female
use WOMEN

Fiche
use MICROFICHE DOCUMENTS

FID (Federation Internationale de Documentation)
  use INTERNATIONAL (GEOGRAPHY)
      INFORMATION SCIENCE SERVICE ORGANIZATIONS

Field (Geographic)
  use GEOGRAPHIC AREAS

Field (Subject)
  use KNOWLEDGE AREAS

File Compression
  use DATA
      TRANSLATION

File Design
  use INTERRELATIONS
      DATA

File Maintenance
  use DATA
      MAINTENANCE

FILE ORGANIZATION
  X Chained Files
    Inverted Files
  BT Interrelations
  PX List Processing

File Security
  use DATA
      SECURITY

Files
  use DATA

Files (Media)
  use STORAGE MEDIA

Filing (Arranging)
  use INTERRELATIONS

Filing (Organizing Storage Media)
  use INTERRELATIONS

FILM (PHOTOGRAPHIC)
  X Microfilm (Medium)
  BT Storage Media
  RT Photography

Reproducing (Of Documents)
Reproducing Equipment
PX APS (Auxiliary Publication Service)
Auxiliary Publication Service (APS)
Filmed Documents
Microfiche (Medium)
Microfilm Readers (Equipment)
Microfilm Systems
National Auxiliary Publications Service

Filmed Documents
use REPRODUCTIONS

Filming
use PHOTOGRAPHY

Filmorex System
use OPTICAL COINCIDENCE TECHNIQUES

FINANCE
X Debts
Depreciation
Expenditures
Funding
Lease
Money
Payment
Rental
Savings
Subsidize
BT Administration
NT Accounting
Budgeting
Costs
Financial Aid
Marketing
RT Advertising
Compensation
Economics
Numeric Processing
Planning
PX Nonprofit Organizations

FINANCIAL AID
BT Finance
NT Grants And Contracts
RT Compensation
PX Fellowships

Findings
  use RESEARCH RESULTS

FINE ARTS
  BT Arts And Humanities
  NT Architecture
     Literature
     Music (Art Form)
     Sculpture
  RT Paintings
     Visual Communication

Firing
  use EMPLOYMENT
     CONTROL FUNCTIONS

FLOODS
  BT Conditions

Floppy Discs
  use MAGNETIC DISCS

Flow (Of Information) (Dissemination Patterns)
  use INFORMATION FLOW

FLOW DIAGRAMS
  BT Graphic Records

FOOD
  BT Substances

Ford Foundation
  use PHILANTHROPIC ORGANIZATIONS

FORECASTS (DOCUMENTS)
  BT Documents
  RT Predicting
     Time

Foreign Language
  use NATURAL LANGUAGES

Foreign Relations
  use INTERNATIONAL RELATIONS

Form (Structure)
  use STRUCTURE

Format Standardization
  use COMPOSING
      CRITERIA

Formatting
  use COMPOSING

FORMS (SURVEY)
  BT Instruments

FORTRAN (PROGRAMMING LANGUAGE)
  BT Artificial Languages

FRANCE
  BT Europe
  PX ADBS (Assoc. Franc. Des Documentalistes Et Des Bibliothecaires Specialises)
      Assoc. Franc. Des Documentalistes Et Des Bibliothecaires Specialises (ADBS)

Free Terms
  use INDEX TERMS (FORMAT AND INTERRELATIONS)

Free-Text Indexing
  use INDEXING
      NATURAL LANGUAGES

FRENCH LANGUAGE
  BT Natural Languages

FREQUENCY
  BT Measures (Measuring Techniques)

Friends
  use FAMILY

Full Text
  use DOCUMENTS

Funding
  use FINANCE

FUTURE
  BT Time

GAME THEORY
  BT Theory
  RT Statistics

General Services Administration (GSA)
  use UNITED STATES GOVERNMENT
      SERVICE ORGANIZATIONS

Genetic Coding System
  use CODING SYSTEMS
      GENETICS

GENETICS
  X Heredity
  BT Biological Sciences
  PX Genetic Coding System

GEOGRAPHIC AREAS
  X Countries
    Field (Geographic)
    Location (Place)
  BT Resources
  NT Africa
    Asia
    Australia
    Canada
    Europe
    International (Geography)
    Latin America
    Mexico
    United States Of America
  RT Cities
    Communities
    Depressed Areas
    National (Nationwide)
    Regional
    Rural
    States
    Suburban
  PX Maps (Geographical)
    Native Speakers
    Vernacular

GEOLOGY
  BT Physical Sciences

GEOMETRY
  BT Mathematics
  NT Topology

GERMAN LANGUAGE
  BT Natural Languages

GERMANY
  BT   Europe

Gestures
  use KINESICS

Ghetto
  use COMMUNITIES

Glossaries
  use DICTIONARIES

GOALS
  X   Objectives
  BT   Environment
  NT   Expectations
       Motivation
       Values
  RT   Resources

GOVERNMENTS
  BT   Organizations
  NT   State Government
       United States Government
       Urban Government
  RT   Control Functions

Graduate Education
  use UNIVERSITY AND COLLEGE EDUCATION

Grammar
  use SYNTAX

Grant Proposals
  use GRANTS AND CONTRACTS
       PROPOSALS

Grants (Of Money)
  use GRANTS AND CONTRACTS

GRANTS AND CONTRACTS
  X   Contracts
       Grants (Of Money)
  BT   Financial Aid
  RT   Laws
  PX   Contract Proposals
       Grant Proposals

Graphic Data
  use DATA
      GRAPHIC RECORDS

Graphic Information
  use DATA
      GRAPHIC RECORDS

GRAPHIC RECORDS
      Note: Used For Diagrammatic And Pictorial Representations.
  X   Charts
      Data, Graphic
      Drawings
      Graphs (Documents)
      Illustrations
      Information, Graphic
      Line Diagrams
      Schematics
      Tables (Tabular Representation)
  BT Audiovisual Software (Documents)
  NT Blueprints
      Flow Diagrams
      Paintings
      Photographs
  RT Algebras
      Audiovisual Production Services
      Data Processing Services
      Numeric Processing
      Records
  PX Computer Graphics
      Engineering Drawings
      Graphic Data
      Graphic Information
      Graphics (Composing)
      Line Plotter
      Maps (Geographical)

Graphics (Composing)
  use COMPOSING
      GRAPHIC RECORDS

GRAPHICS PRINTING EQUIPMENT
  X   Plotter (Equipment)
  BT Printing Equipment

Graphs (Documents)
  use GRAPHIC RECORDS

Gray Codes
  use CODING SYSTEMS

Great Britain
  use UNITED KINGDOM

Group Behavior
  use GROUPS
    BEHAVIOR

Group Communication
  use GROUPS
    COMMUNICATION

Grouping (Of Documents)
  use DOCUMENTS
    INTERRELATIONS

GROUPS
  X Clubs (Social)
    Committees
  BT Human Beings
  NT Cultures
    Disadvantaged Persons
    Ethnic Groups
    Handicapped Persons
    Socioeconomic Groups
  PX Discussion
    Group Behavior
    Group Communication
    Membership (In A Group)

GROWTH
  X Development (Physical)
  BT Change

GSA (General Services Administration)
  use UNITED STATES GOVERNMENT
    SERVICE ORGANIZATIONS

Guides (Reference Books)
  use BOOKS

Guides (Rules)
  use RULES

Handbooks
  use BOOKS

HANDICAPPED PERSONS
  X   Blind Persons
       Deaf Persons
       Disabled Persons
       Immobilized Persons
       Mental Disability
  BT   Groups
  RT   Disadvantaged Persons
       Dysfunctions
  PX   Special Education

Handwriting
  use WRITING

Hardcopy
  use OUTPUT RECORDS

Hardware
  use EQUIPMENT

Headings, Subject
  use INDEX TERMS (DOCUMENT SURROGATES)

HEALTH CARE
  X   Medical Care
       Mental Health Care
       Nursing
  BT   Biomedical Sciences
  NT   Dentistry
  PX   Computer Assisted Medical Diagnosis
       Health Professionals

Health Professionals
  use HEALTH CARE
       PERSONNEL

HEALTH, EDUCATION AND WELFARE DEPARTMENT (HEW)
  X   HEW (Health, Education And Welfare
            Department)
  BT   United States Government
  NT   Public Health Service

HEARING
  BT   Senses

Heredity
  use GENETICS

HEW (Health, Education And Welfare Department)
  use HEALTH, EDUCATION AND WELFARE DEPARTMENT (HEW)

Hierarchy
  use CLASSIFICATION SYSTEMS

HIGH SCHOOL EDUCATION
  X Secondary Education
  BT Education
  RT Courses (Educational)
      Curricula

Higher Education
  use UNIVERSITY AND COLLEGE EDUCATION

Hiring
  use EMPLOYMENT
      CONTROL FUNCTIONS

Historical Documents
  use DOCUMENTS
      HISTORY

Historical Research
  use ARCHIVAL RESEARCH

HISTORY
  X Evolution
  BT Behavioral Sciences
  PX Historical Documents

Hollerith Cards
  use PUNCHED CARDS

HOLOGRAMS
  BT Photographs
  RT Reproducing (Of Documents)

HOLOGRAPHY
  BT Photography

Honorary Degrees
  use AWARDS
      ACADEMIC ORGANIZATIONS

Hospital Information Systems
  use HOSPITALS
      STORAGE AND RETRIEVAL SYSTEMS

HOSPITALS
  X   Mental Hospitals
  BT   Buildings
  PX   Hospital Information Systems

House Organs
  use PUBLICATIONS (PERSONAL OR ORGANIZATIONAL)

Houses
  use RESIDENTIAL FACILITIES

Human Behavior
  use BEHAVIOR

HUMAN BEINGS
  BT   Population
  NT   Groups
       Men
       Women
  RT   Cybernetics
       Ethics
       Human Relations
       Human Resources
       Interaction
       Neurology
  PX   Engineering, Human Factor
       Human Communication
       Human Engineering
       Human Memory

Human Communication
  use COMMUNICATION
       HUMAN BEINGS

Human Engineering
  use ENGINEERING
       HUMAN BEINGS

Human Memory
  use MEMORY
       HUMAN BEINGS

HUMAN RELATIONS
  X   Integration (Social)
       Involvement
       Public Relations
       Publicity
  BT   Relations

NT Interpersonal Relations
Satisfaction
RT Behavior
Consulting Services
Human Beings
Interaction
Personnel
PX Professional Behavior

HUMAN RESOURCES
X Manpower
BT Resources
NT Ability
Humor
Intellect (Intelligence)
Knowledge (State Of Being Informed)
Personality
RT Human Beings
Opportunity

Humanities
use ARTS AND HUMANITIES

HUMOR
BT Human Resources
PX Humorous Documents

Humorous Documents
use DOCUMENTS
HUMOR

HUNGARY
BT Europe

HYBRID COMPUTERS
X Digital-analog Computers
BT Computers
RT Analog Equipment
Automata
Electronic Digital Computers

Hypotheses
use THEORY

Hypothesis Testing
use THEORY
RESEARCH

IBM Cards
  use PUNCHED CARDS

ICSU (International Council of Scientific Unions)
  use SCIENCE AND TECHNOLOGY
      SERVICE ORGANIZATIONS
      INTERNATIONAL (GEOGRAPHY)

Ideas
  use CONCEPTS

IDENTIFICATION
  X Definition
  BT Control Functions

Identification (Documents Of)
  BT Documents

Idioms
  use NATURAL LANGUAGES

IEEE (Institute Of Electrical And Electronic Engineers)
  use ENGINEERING
      PROFESSIONAL ORGANIZATIONS

IFIP (International Federation Of Information Processing)
  use INFORMATION SCIENCE SERVICE ORGANIZATIONS
      INTERNATIONAL (GEOGRAPHY)

IFLA (International Federation Of Library Associations)
  use INFORMATION SCIENCE SERVICE ORGANIZATIONS
      INTERNATIONAL (GEOGRAPHY)

IIA (Information Industry Association)
  use INFORMATION SCIENCE SERVICE ORGANIZATIONS
      MANUFACTURING AND SALES ORGANIZATIONS

ILL (Interlibrary Loan)
  use INTERLIBRARY LOAN

Illiteracy
  use ABILITY
      READING

Illumination
  use LIGHT

Illustrations
  use GRAPHIC RECORDS

Illustrations (Photographic)
  use PHOTOGRAPHS

ILR (Institute Of Library Research)
  use LIBRARY SCIENCE
  RESEARCH

Image-Making (Of Documents)
  use REPRODUCING (OF DOCUMENTS)

Images, Document
  use REPRODUCTIONS

Imagination
  use CREATIVITY

Immobilized Persons
  use HANDICAPPED PERSONS

Impact
  use RELATIONS

Impede
  use PREVENTION

IMPROVEMENT
  BT Interaction

Incapability
  use CAPACITY

Incentives
  use MOTIVATION

INCIDENCE
  BT Measures (Measuring Techniques)

Income
  use WAGES

INDEPENDENCE
  BT Relations

NT Privacy
RT Structure

Independent Variable
use VARIABLES

Index (Evaluation Standard)
use INDICATORS

Index Instructions (To Users Of An Index)
use INDEXES
INSTRUCTING
USERS (OF SERVICES)

Index Medicus
use BIOMEDICAL SCIENCES
INDEXES

Index Performance
use INDEXES
PERFORMANCE

INDEX TERMS (DOCUMENT SURROGATES)
X Aspect (Indexing Term)
Attribute (Indexing Term)
Call Number
Characteristic (Indexing)
Descriptors
Document Surrogates
Entry (Indexing)
Headings, Subject
Keywords
Subjects (Of Documents)
Terms (Indexing)
BT Indicators
RT Concepts
Thesauri
PX Index Terms, Weighting

INDEX TERMS (FORMAT AND INTERRELATIONS)
X Added Entry
Bound Terms
Cross Referencing
Dictionary Catalog
Free Terms
Interfixes
Links (Indexing)
Main Entry

Roles (Indexing)
See Also Reference
See Reference
Tracings (Indexing)
BT Structure
RT Interrelations

Index Terms, Weighting
use INDEX TERMS (DOCUMENT SURROGATES)
WEIGHTING

Indexer Consistency
use INDEXING
CONSISTENCY

Indexers
use INDEXING
INFORMATION SCIENCE PERSONNEL

INDEXES
X Alphabetic Subject Code
Concordances
Contents Pages (Tables Of Contents Of Documents)
Directories
Subject Catalog
BT Documents
NT Citation Indexes
KWIC Indexes
RT Bibliographies
PX AIM (Abridged Index Medicus)
Articulated Indexes
Back-Of-The-Book Indexes
Bibliographic Data Bases
Book Catalogs
Card Catalogs
Cards, Subject
Catalogs
Catalogs, Union
CBAC (Chemical-Biological Activities) (Index)
Chemical Abstracts (Publication)
Classified Index
Conventional Systems (Of Indexing)
Cumulated Indexes
Data Bases (For Automated Storage And Retrieval)
Excerpta Medica

Index Instructions (To Users Of An Index)
Index Medicus
Index Performance
Information Collecting (Use Of Indexes)
Natural Language Indexes
Personal Indexes (Of Scientists)
Shelf List
Subject Cards
Union Catalogs
Union Lists

INDEXING
X Analysis, Document
Analytical Subject Entry
Document Description
Document Representation
Document Surrogation
Indexing, Subject
Labeling
Subject Analysis (Of Documents)
BT Bibliography (Process)
NT Coordinate Indexing
KWIC Indexing
RT Evaluation Techniques
Faceted Classification Systems
Information
PX AACR (Anglo-American Cataloging Rules)
Anglo-American Cataloging Rules (AACR)
Author Indexing
Automated Indexing
Automated Subject Analysis
Cataloging, Descriptive
Chemical Indexing
Classificationists
Classifier
Computer Assisted Indexing
Cranfield Project
Descriptive Cataloging (Process)
Free-Text Indexing
Indexer Consistency
Indexers
Indexing Instructions (To Data Processing Machines)
Indexing Instructions (To Human Indexers)
Indexing Services
Indexing Techniques
Indexing Theory
Indexing, Phonetic

National Federation Of Abstracting And Indexing Services
Phonetic Indexing
Source Indexing

Indexing Instructions (To Data Processing Machines)
use INDEXING
PROGRAMS

Indexing Instructions (To Human Indexers)
use INDEXING
INFORMATION SCIENCE PERSONNEL
INSTRUCTING

Indexing Languages
use CLASSIFICATION SYSTEMS

Indexing Services
use INDEXING
INFORMATION SCIENCE SERVICE ORGANIZATIONS

Indexing Techniques
use INDEXING
TECHNIQUES

Indexing Theory
use INDEXING
THEORY

Indexing, Phonetic
use INDEXING
SOUND

Indexing, Subject
use INDEXING

INDIA
BT Asia
PX Indian National Scientific Documentation Center (INSDOC)
INSDOC (Indian National Scientific Documentation Center)

Indian National Scientific Documentation Center (INSDOC)
use INDIA
SCIENCE AND TECHNOLOGY
INFORMATION SCIENCE SERVICE ORGANIZATIONS

INDICATORS
BT Criteria
NT Index Terms (Document Surrogates)
RT Abstracts (Of Documents)
Statistics

Industrial Information Systems
use MANUFACTURING AND SALES ORGANIZATIONS
STORAGE AND RETRIEVAL SYSTEMS

Ineligibility
use RULES

Inexpensive
use COSTS

INFLUENCING
BT Interaction

INFORMATION
Note: Use More Specific Or Precise Terms Instead, Whenever Possible; See The Related Terms For Examples.
BT Input
NT Data
Documents
RT Bibliography (Process)
Data Processing
Indexing
Information Science
Information Science Service Organizations
Input Records
Knowledge (State Of Being Informed)
Libraries
Service Organizations
Storage And Retrieval Processes
PX Information (Structure Of)
Information Industry
Information Resources
Information Transfer
Multidisciplinary Information
Science Information

Information (Structure Of)
use INFORMATION
STRUCTURE

Information Centers
use INFORMATION SCIENCE SERVICE ORGANIZATIONS

Information Coding
  use CODING
      DATA

Information Collecting (Data)
  use DATA GATHERING

Information Collecting (Documents)
  use ACQUIRING
      DOCUMENTS

Information Collecting (Use Of Indexes)
  use SEARCHING
      INDEXES

Information Collection (Noun) (Group Of Documents)
  use DOCUMENTS
      STORAGE AND RETRIEVAL SYSTEMS

Information Collection (Noun) (Pool Of Data)
  use DATA
      STORAGE AND RETRIEVAL SYSTEMS

Information Crisis
  use USES (OF RESOURCES)
      DYSFUNCTIONS

Information Dissemination
  use DISSEMINATING

Information Exchange (Data)
  use DATA
      EXCHANGING

Information Exchange (Documents)
  use DOCUMENTS
      EXCHANGING

Information Exchange (Interpersonal)
  use COMMUNICATION
      INTERPERSONAL RELATIONS

INFORMATION FLOW
      Note: Used For Patterns Of Communication Activity Indigenous To Particular Geographic or Knowledge Areas.
  X   Channels Of Communication
      Flow (Of Information) (Dissemination Patterns)

Invisible Colleges
Metabolism Of Information
BT Disseminating
RT SDI
Transmission
PX ASCA (Automatic Subject Citation Alert) (Tradename)
Automatic Subject Citation Alert (ASCA) (Tradename)
Computer Assisted Literature Alerting
Crisis, Information
Current Awareness

Information Gathering
use DATA GATHERING

Information Handling Equipment
use DATA PROCESSING EQUIPMENT

Information Industry
use INFORMATION

Information Industry Association (IIA)
use MANUFACTURING AND SALES ORGANIZATIONS
INFORMATION SCIENCE SERVICE ORGANIZATIONS

Information Management
use BIBLIOGRAPHY (PROCESS)

INFORMATION NEED
BT Need
RT Inquiries
Responses (To Inquiries)
PX Information Seeking Behavior

Information Networks (Automated)
use TIME-SHARING SERVICES
INFORMATION SCIENCE SERVICE ORGANIZATIONS

Information Officers
use INFORMATION SCIENCE PERSONNEL

Information Processing
use DATA PROCESSING

Information Resources
use INFORMATION
RESOURCES

Information Retrieval (Process)
use STORAGE AND RETRIEVAL PROCESSES

Information Retrieval (Systems)
use STORAGE AND RETRIEVAL SYSTEMS

INFORMATION SCIENCE
X Informatology
NT Environment
Feedback And Control
Input
Output
Processes (Processing)
RT Data Processing
Information
Library Science
PX Advanced Research Projects Agency (ARPA)
American Documentation (Journal)
Annual Review Of Information Science And Technology (ARIST)
ARIST (Annual Review Of Information Science And Technology)
ARPA (Advanced Research Projects Agency)
Arpanet
Communication Theory
Documentation Literature
International Federation For Documentation (FID)

INFORMATION SCIENCE PERSONNEL
X Documentalists
Information Officers
Information Scientists
BT Personnel
NT Users (Of Resources)
PX Abstracter
Chemists, Literature
Classificationists
Classifier
Editors
Educators
Indexers
Indexing Instructions (To Human Indexers)
Instructors
Interpreter (Person)
Literature Chemists
Professors
Teachers

Translater (Human)
Transmitters (Human)

INFORMATION SCIENCE PROFESSIONAL ORGANIZATIONS
X ADI (American Documentation Institute)
American Documentation Institute (ADI)
American Society For Information Science (ASIS)
ASIS (American Society For Information Science)
BT Professional Organizations
PX ACM (Association For Computing Machinery)
ADBS (Assoc. Franc. Des Documentalistes Et Des Bibliothecaires Specialises)
AFIPS (American Federation Of Information Processing Societies)
ALA (American Library Association)
American Federation Of Information Processing Societies (AFIPS)
American Library Association (ALA)
ARL (Association Of Research Libraries)
ASLIB (Association Of Library And Information Bureaux)
Assoc. Franc. Des Documentalistes Et Des Bibliothecaires Specialises (ADBS)
Association For Computer Machinery (ACM)
Association Of Library And Information Bureaux (ASLIB)
Association Of Research Libraries (ARL)
CBE (Conference Of Biological Editors)
Conference Of Biological Editors (CBE)
Medical Library Association (MLA)
MLA (Medical Library Association)
SLA (Special Libraries Association)
Special Libraries Association (SLA)

INFORMATION SCIENCE SERVICE ORGANIZATIONS
X Audiovisual Centers
Bibliographic Centers
Centralized Information Centers
Clearinghouse (information)
Documentation Centers
Information Centers
Information Services (Organizations)
Literature Retrieval Centers
BT Service Organizations
NT Storage And Retrieval Systems
Time-Sharing Services

RT Electronic Digital Computers
Information
Library Science

PX Abstracting Services
ACM (Association For Computing Machinery)
ADBS (Assoc. Franc. Des Documentalistes Et Des Bibliothecaires Specialise)
AFIPS (American Federation Of Information Processing Societies)
Air Pollution Information Center (APTIC)
APS (Auxiliary Publication Service)
APTIC (Air Pollution Information Center)
Auxiliary Publication Service (APS)
Biosciences Information Service (BIOSIS) (Tradename)
BIOSIS (Biosciences Information Service) (Tradename)
CFSTI (Clearinghouse For Federal Scientific And Technical Information)
Clearinghouse For Federal Scientific And Technical Information (CFSTI)
Committee Z39
Educational Resources Information Center (ERIC)
ERIC (Educational Resources Information Center)
Federation Internationale de Documentation (FID)
FID (Federation Internationale de Documentation)
IFIP (International Federation Of Information Processing)
IFLA (International Federation of Library Associations)
IIA (Information Industry Association)
Information Industry Association (IIA)
Indexing Services
Indian National Scientific Documentation Center (INSDOC)
Information Networks (Automated)
INSDOC (Indian National Scientific Documentation Center)
International Federation of Information Processing (IFIP)
International Federation of Library Associations (IFLA)
Japan Information Center of Science And Technology (JICST)

JICST (Japan Information Center of
Science And Technology)
Learning Resource Centers
Library Standards (For Services)
National Federation Of Abstracting And
Indexing Services
National Technical Information Service
(NTIS)
NTIS (National Technical Information
Service)
Office Of Scientific Information Services
(OSIS)
OSIS (Office Of Scientific Information
Services)
Science Information Exchange (SIE)
SIE (Science Information Exchange)
Z39 Committee

Information Scientists
use INFORMATION SCIENCE PERSONNEL

Information Searching
use SEARCHING

Information Searching Systems
use STORAGE AND RETRIEVAL SYSTEMS

Information Seeking Behavior
use INFORMATION NEED
BEHAVIOR

Information Services (Organizations)
use INFORMATION SCIENCE SERVICE ORGANIZATIONS

Information Services (Responses To Inquiries)
use RESPONSES (TO INQUIRIES)

Information Systems
use STORAGE AND RETRIEVAL SYSTEMS

INFORMATION THEORY
BT Theory

Information Transfer
use INFORMATION
TRANSMISSION

Information Users (Authors, Service Providers)
use USERS (OF RESOURCES)

Information Users (Information System Clientele)
  use USERS (OF SERVICES)

Information Uses
  use USES (OF RESOURCES)

Information, Graphic
  use GRAPHIC RECORDS

Information, Science
  use SCIENCE AND TECHNOLOGY

Informatology
  use INFORMATION SCIENCE

Innovation
  use CHANGE

INPUT
  BT Information Science
  NT Equipment
      Information
      Input Process
      Libraries
      Museums
      Need
      Organizations
      Techniques
  RT Personnel

Input Data
  use DATA

INPUT PROCESS
  X Keying
  BT Input
  NT Acquiring
      Inquiries
      Stimuli
  RT Input Records
      Input-Output Equipment
      Reading

INPUT RECORDS
      Note: Used For Media Or Data On Which One
            Or More Processing Steps Will Be
            Performed.
  X Data, Input

BT Records
RT Data
Data Processing
Decision-Making
Documents
Information
Input Process
Input-Output Equipment
Inquiries
Output Records
Programs

INPUT-OUTPUT EQUIPMENT
X Acoustic Couplers
Channel (Hardware)
Communication Equipment
Computer Input Devices
Computer Output Devices
Display Devices (Data Processing)
Keying Equipment
Magnetic (Card, Tape, Disc) Typewriters
Punching Equipment
Reader-Printer
Tape Typewriters
Teletypewriters
BT Data Processing Equipment
NT Audiovisual Equipment
Computer Terminals
Light Pen
Printing Equipment
Reading Equipment
Receivers (Equipment)
Reproducing Equipment
Telephone
Transmitters (Equipment)
RT Electromechanical Data Processing Equipment
Input Process
Input Records
Punched Cards
Techniques
Time-Sharing
Word Processing
PX Audio Input-Output
Automated Reading
Automated Speech
Line Plotter

INQUIRIES
  X   Queries
       Questions
       Reference Questions
  BT   Input Process
  RT   Evaluation
       Feedback And Control
       Information Need
       Input Records
  PX   Query Formulation

INQUIRY NEGOTIATION
  BT   Storage And Retrieval Processes

INSDOC (Indian National Scientific Documentation Center)
  use INDIA
       SCIENCE AND TECHNOLOGY
       INFORMATION SCIENCE SERVICE ORGANIZATIONS

Institute Of Electrical And Electronic Engineers (IEEE)
  use ENGINEERING
       PROFESSIONAL ORGANIZATIONS

Institute Of Library Research (ILR)
  use LIBRARY SCIENCE
       RESEARCH

Institutes (Communicating Via Meetings)
  use MEETINGS (COMMUNICATING VIA MEETINGS)

Institutions
  use ORGANIZATIONS

INSTRUCTING
  X   Reeducation
       Teaching
       Training
  BT   Communication
  NT   Computer Assisted Instructing (CAI)
       Courses (Educational)
       Curricula
       Self-Instructing
  RT   Communication
       Education
       Learning
       Multimedia Techniques

Postgraduate Education
Programs
Techniques
PX Automated Teaching
Consumer Education
Educational Television (ETV)
Educators
ETV (Educational Television)
Faculty
Index Instructions (To Users Of An Index)
Indexing Instructions (To Human Indexers)
Instructors
Language Training
Learning Machines
Medical Education
Peace Corps
Professors
Programmed Instruction
Special Education
Syllabus
Teachers
Tutorial Documents

Instruction (Receiving Of)
use LEARNING

Instructions (Algorithmic)
use PROGRAMS

Instructors
use INSTRUCTING
INFORMATION SCIENCE PERSONNEL

INSTRUMENTS
BT Equipment
NT Forms (Survey)
Questionnaires

INSURANCE (LIFE)
BT Compensation
RT Economics

Insurance (Subject Area)
use ECONOMICS

Integration (Social)
use HUMAN RELATIONS

INTELLECT (INTELLIGENCE)
X Mind
Thinking
BT Human Resources
NT Cognition
Learning
Reasoning
Recognition
RT Associating
Knowledge (State Of Being Informed)
Memory
Neurology
PX Artificial Intelligence
IQ (Intelligence Quotient)

Intelligent Computer Terminals
use MICROPROCESSORS

Inter-University Communications Council
use TIME-SHARING SERVICES
ACADEMIC ORGANIZATIONS

INTERACTION
X Causes
Determinants
Effects
Liaison
Masking
Participation
Teamwork
BT Administration
NT Accessing
Adjustment
Availability
Browsability
Compatibility
Convertibility
Cooperation
Coordination
Improvement
Influencing
Man-Machine Interaction
Redundancy
Risk
Security
Stress
RT Effectiveness
Efficiency

Human Beings
Human Relations
International (Geography)
International Relations
Interrelations
Structure

Interactive Computer Services
use TIME-SHARING SERVICES

Interactive Computer Use (Man-Machine)
use MAN-MACHINE INTERACTION

Intercitations
use CITATIONS

Interest Profiles
use INTERESTS

INTERESTS
X Interest Profiles
Preference
Profiles (Of Interests)
BT Motivation

Interface
use INTERRELATIONS

Interfixes
use INDEX TERMS (FORMAT AND INTERRELATIONS)

INTERLIBRARY LOAN
X ILL (Interlibrary Loan)
BT Exchanging

INTERLINGUA
BT Artificial Languages

Intermediate Languages
use ARTIFICIAL LANGUAGES

Internal Reports
use ORGANIZATIONS
RECORDS

INTERNATIONAL (GEOGRAPHY)
Note: Used To Express Intercountry Activities.

BT Geographic Areas
RT Interaction
Organizations
Regional
PX Federation Internationale de Documentation (FID)
FID (Federation Internationale de Documentation)
ICSU (International Council Of Scientific Unions)
IFIP (International Federation Of Information Processing)
IFLA (International Federation Of Library Associations)
International Council Of Scientific Unions (ICSU)
International Federation For Documentation (FID)
International Federation Of Information Processing (IFIP)
International Federation Of Library Associations (IFLA)
ISO (International Standards Organization)
International Standards Organization (ISO)
Peace Corps
Unisist

International Council Of Scientific Unions (ICSU)
use SCIENCE AND TECHNOLOGY
SERVICE ORGANIZATIONS
INTERNATIONAL (GEOGRAPHY)

International Federation For Documentation (FID)
use INFORMATION SCIENCE
PROFESSIONAL ORGANIZATIONS
INTERNATIONAL (GEOGRAPHY)

International Federation Of Information Processing (IFIP)
use DATA PROCESSING
INFORMATION SCIENCE SERVICE ORGANIZATIONS
INTERNATIONAL (GEOGRAPHY)

International Federation Of Library Associations (IFLA)
use INFORMATION SCIENCE SERVICE ORGANIZATIONS
INTERNATIONAL (GEOGRAPHY)

INTERNATIONAL RELATIONS
  X   Foreign Relations
  BT   Relations
  RT   Interaction

International Standards Organization (ISO)
  use CRITERIA
       INTERNATIONAL (GEOGRAPHY)
       SERVICE ORGANIZATIONS

Internship
  use POSTGRADUATE EDUCATION

Interpersonal Communication
  use INTERRELATIONS
       COMMUNICATION

INTERPERSONAL RELATIONS
  BT   Human Relations
  PX   Information Exchange (Interpersonal)
       Membership (In A Group)

Interpreter (Equipment)
  use TRANSLATION
       EQUIPMENT

Interpreter (Person)
  use LANGUAGE TRANSLATION
       INFORMATION SCIENCE PERSONNEL

INTERPRETING
  BT   Translation
  PX   Photointerpretation

INTERRELATIONS
  X   Arranging
       Centralizing
       Decentralization
       Filing (Arranging)
       Filing (Organizing Storage Media)
       Isolation
       Ordering (Arranging)
       Organization
       Organizing
       Patterns
  BT   Administration
  NT   Associating
       Bias

Collating
Comparing
Composing
Consistency
Constructing
Extracting
File Organization
Meeting Organization
Merging
Opportunity
Permutation
Ranking
Readiness
Sequencing
Sorting
Structure
RT Change
Classification Systems
Coding Systems
Data Analysis
Electromechanical Data Processing Equipment
Index Terms (Format And Interrelations)
Interaction
Randomness
Relations
PX Commissions (Administrative Groups)
Data Description Languages
Document Organization (Process)
File Design
Grouping (Of Documents)
Many-To-One Communication
One-To-Many Communication
Query Formulation
Random Access Storage
Search Strategy

INTERVIEWING
BT Surveying
RT Data
Questionnaire Techniques

Inventorying
use ACCOUNTING

Inverted Files
use FILE ORGANIZATION

Investigating
use EVALUATION

Investigators
  use EVALUATION
      PERSONNEL

Invisible Colleges
  use INFORMATION FLOW

Involvement
  use HUMAN RELATIONS

IQ (Intelligence Quotient)
  use MEASURES (MEASURING TECHNIQUES)
      INTELLECT (INTELLIGENCE)

IRELAND
  BT Europe

ISO (International Standards Organization)
  use CRITERIA
      INTERNATIONAL (GEOGRAPHY)
      SERVICE ORGANIZATIONS

Isolation
  use INTERRELATIONS

ISRAEL
  BT Asia

ITALY
  BT Europe

Iterative Search Processes
  use SEARCHING

Itinerants
  use MIGRANTS

J

JAPAN
  BT Asia
  PX Japan Information Center Of Science And Technology (JICST)
      JICST (Japan Information Center Of Science And Technology)

Japan Information Center Of Science And Technology (JICST)
use JAPAN
SCIENCE AND TECHNOLOGY
INFORMATION SCIENCE SERVICE ORGANIZATIONS

JAPANESE LANGUAGE
BT Natural Languages

JICST (Japan Information Center Of Science And Technology)
use JAPAN
SCIENCE AND TECHNOLOGY
INFORMATION SCIENCE SERVICE ORGANIZATIONS

Job Analysis
use EMPLOYMENT
SPECIFICATIONS

Job Description
use EMPLOYMENT
SPECIFICATIONS

Job Performance
use PERFORMANCE

Job Satisfaction
use SATISFACTION

Jobs
use EMPLOYMENT

Judging
use DECISION-MAKING

K

Keying
use INPUT PROCESS

Keying Equipment
use INPUT-OUTPUT EQUIPMENT

Keyword-In-Context Indexes
use KWIC INDEXES

Keyword-In-Context Indexing
  use KWIC INDEXING

Keyword-Out-Of-Context Indexes
  use KWIC INDEXES
      ARRANGEMENT

Keywords
  use INDEX TERMS (DOCUMENT SURROGATES)

Kinescopes
  use RECORDS
      TELEVISION

KINESICS
  X Gestures
  BT Symbol Sets
  RT Body Language
      Linguistics
      Signals
      Visual Communication

KNOWLEDGE (STATE OF BEING INFORMED)
  X Understanding
  BT Human Resources
  RT Data
      Information
      Intellect (Intelligence)
      Memory

KNOWLEDGE AREAS
  X Disciplines
      Field (Subject)
  BT Resources
  NT Arts And Humanities
      Mathematics
      Philosophy
      Science And Technology
  PX Multidisciplinary Information

KOREA
  BT Asia

KWIC INDEXES
  X Keyword-In-Context Indexes
  BT Indexes
  RT KWIC Indexing
  PX Keyword-Out-Of-Context Indexes
      KWOC Indexes

KWIC INDEXING
  X   Keyword-In-Context Indexing
  BT   Indexing
  RT   KWIC Indexes
       Permutation

KWOC Indexes
  use KWIC INDEXES
       ARRANGEMENT

L

Labeling
  use INDEXING

Labor Unions
  use EMPLOYMENT
       ORGANIZATIONS

Laboratories
  use FACILITIES

Laboratory Data
  use DATA

Laboratory Records
  use RECORDS

Lag
  use TIME

Language Barriers
  use LANGUAGES
       BARRIERS

Language Problems
  use LANGUAGES
       DYSFUNCTIONS

Language Processing (Computers)
  use DATA PROCESSING
       COMPUTER APPLICATIONS

Language Structures
  use LANGUAGES
       STRUCTURE

Language Symbols
  use SYMBOL SETS

Language Training
  use NATURAL LANGUAGES
      INSTRUCTING

LANGUAGE TRANSLATION
  BT Translation
  RT Languages
      Word Processing
  PX Automated Language Translation
      Interpreter (Person)
      Language Translation (Automated)
      Machine Translation
      Translater (Human)

Language Translation (Automated)
  use COMPUTER APPLICATIONS
      LANGUAGE TRANSLATION

LANGUAGE TRANSLATION SERVICES
  BT Delivery Of Services

LANGUAGES
  X Metalanguage
      Multilingual
  BT Resources
  NT Artificial Languages
      Natural Languages
      Symbol Sets
  RT Language Translation
      Linguistics
  PX Barriers, Language
      Language Barriers
      Language Problems
      Language Structures
      Words

LASERS
  BT Equipment

LATIN AMERICA
  BT Geographic Areas

Lattices
  use STRUCTURE
      ALGEBRAS

LAW
  X   Legal
       Legislation
       Litigation
  BT   Behavioral Sciences
  RT   Bylaws
       Rules
  PX   ABA (American Bar Association)
       American Bar Association (ABA)
       Attorneys
       Automated Law Enforcement
       FBI (Federal Bureau Of Investigation)
       FCC (Federal Communications Commission)
       Federal Bureau Of Investigation (FBI)
       Federal Communications Commission (FCC)
       Law Enforcement Information Systems
       Lawyers
       Legal Research (Searching Legal Literature)
       Police

Law Enforcement Information Systems
  use LAW
       STORAGE AND RETRIEVAL SYSTEMS

LAWS
  BT   Rules
  NT   Bylaws
       Copyright Law
       Courts
  RT   Grants And Contracts

Lawyers
  use LAW
       PERSONNEL

LAY PUBLICATIONS
  BT   Documents
  RT   Advertising Literature
       News Publications
       Publications (Personal Or Organizational)

Leadership
  use ABILITY

LEARNING
  X   Instruction (Receiving Of)
  BT   Intellect (Intelligence)
  RT   Cognition

Concepts
Education
Instructing
Multimedia Techniques
Postgraduate Education
Recognition
PX Adult Education
Learning (Computer-Assisted)
Learning Problems
Programmed Learning
Students
Trainees
Work Study Program

Learning (Computer-Assisted)
use MAN-MACHINE INTERACTION
LEARNING

Learning Machines
use INSTRUCTING
EQUIPMENT

Learning Problems
use LEARNING
DYSFUNCTIONS

Learning Resource Centers
use MULTIMEDIA TECHNIQUES
INFORMATION SCIENCE SERVICE ORGANIZATIONS

LEARNING THEORY
BT Theory

Lease
use FINANCE

Legal
use LAW

Legal Research (Searching Legal Literature)
use LAW
SEARCHING

Legal Rights
use RIGHTS

Legislation
use LAW

Leisure
  use TIME

Length (Spatial Dimension)
  use SIZE

Length Of Time
  use TIME

LENSES
  BT Components

Lessons
  use COURSES (EDUCATIONAL)

Letters (Correspondence)
  use CORRESPONDENCE (DOCUMENTS)

Letters (Symbols)
  use SYMBOL SETS

Lexicographical Analysis
  use LINGUISTICS

Lexicon (Controlled Language)
  use THESAURI

Lexicon (Natural Language)
  use DICTIONARIES

Liability
  use RESPONSIBILITY

Liaison
  use INTERACTION

Librarians
  use LIBRARY SCIENCE
  PERSONNEL

LIBRARIES
  Note: Reserved For Facilities Providing Services Of Professional Librarians; To Refer To Use Of Personal Document Collections Use: "Documents," "Storage And Retrieval Processes."
  X Archives (Organized)
  Public Libraries

BT Input
NT Library Of Congress
National Agricultural Library
National Library Of Medicine
RT Academic Organizations
Buildings
Data
Defense Documentation Center (DDC)
Information
Museums
Self-Instructing
Service Organizations
United States Government
PX Academic Libraries
BALLOTS (Bibliographic Automation Of Large Libraries Using Time-Sharing)
Cooperative Library Programs
Library Administration
Library Assistants
Library Networks
Library Operations
Library Research (Evaluation Of Library Functions)
Library Resources
Library Standards (Building Standards)
Library Technology
Regional Medical Libraries
Research Libraries
SLA (Special Libraries Association)
Special Libraries Association (SLA)
University Libraries

Library Administration
use LIBRARIES
ADMINISTRATION

Library Assistants
use PARAPROFESSIONALS
LIBRARIES

LIBRARY CLASSIFICATION SYSTEMS
BT Classification Systems
NT Dewey Decimal Classification System
Library Of Congress Classification System
Universal Decimal Classification System
PX Shelf List

Library Networks

use LIBRARIES
EXCHANGING

LIBRARY OF CONGRESS
BT Libraries

LIBRARY OF CONGRESS CLASSIFICATION SYSTEM
BT Library Classification Systems

Library Operations
use LIBRARIES
NORMAL FUNCTIONS

Library Research (Evaluation Of Library Functions)
use LIBRARIES
OPERATIONS RESEARCH

Library Research (Literature Searches)
use SEARCHING

Library Resources
use LIBRARIES
RESOURCES

Library Schools
use LIBRARY SCIENCE
ACADEMIC ORGANIZATIONS

LIBRARY SCIENCE
BT Behavioral Sciences
NT Bibliography (Process)
RT Delivery Of Services
Information Science
Information Science Service Organizations
PX ALA (American Library Association)
American Library Association (ALA)
ARL (Association Of Research Libraries)
Association Of Research Libraries (ARL)
ILR (Institute Of Library Research)
Institute Of Library Research (ILR)
Librarians
Library Schools
Library Standards (For Services)

Library Science Techniques
use DATA PROCESSING TECHNIQUES

Library Services

use DELIVERY OF SERVICES

Library Standards (Building Standards)
use LIBRARIES
BUILDINGS
CRITERIA

Library Standards (For Services)
use LIBRARY SCIENCE
CRITERIA
INFORMATION SCIENCE SERVICE ORGANIZATIONS

Library Technology
use LIBRARIES
EQUIPMENT

Licensing
use CONTROL FUNCTIONS

Life Style
use BEHAVIOR

LIGHT
X Illumination
BT Resources
RT Senses
Physical Sciences

LIGHT PEN
BT Input-Output Equipment

Line Diagrams
use GRAPHIC RECORDS

Line Plotter
use GRAPHIC RECORDS
INPUT-OUTPUT EQUIPMENT

Line Printers
use PRINTING EQUIPMENT

Linguistic Analysis
use LINGUISTICS

LINGUISTICS
X Analysis, Linguistic
Applied Linguistics
Lexicographical Analysis

Linguistic Analysis
BT Behavioral Sciences
NT Semantics
Syntax
RT Behavior
Body Language
Coding Systems
Computer Applications
Evaluation Techniques
Kinesics
Languages
Logic
Speech
Symbol Sets
Symbolic Logic
Word Processing
PX Automated Language Processing
Computational Linguistics
Content Analysis
Psycholinguistics

Links (Indexing)
use INDEX TERMS (FORMAT AND INTERRELATIONS)

LISP (PROGRAMMING LANGUAGE)
BT Artificial Languages

List Processing
use DATA PROCESSING
FILE ORGANIZATION

Literacy
use ABILITY
READING

Literary Research (Searching The Literature)
use SEARCHING

LITERATURE
Note: Use "Documents" Instead Except When "Literature" Is Used In The Abstract Sense.
BT Fine Arts

Literature Chemists
use INFORMATION SCIENCE PERSONNEL
CHEMISTRY

Literature Retrieval Centers
use INFORMATION SCIENCE SERVICE ORGANIZATIONS

Litigation
use LAW

Living Conditions
use CONDITIONS

Living Facilities
use FACILITIES

Loan (Of Library Materials)
use DELIVERY OF DOCUMENTS

Location (Organizational Affiliation)
use ORGANIZATIONS

Location (Place)
use GEOGRAPHIC AREAS

LOGIC
X Conjunctions (Logical)
BT Mathematics
NT Symbolic Logic
RT Algebras
Concepts
Linguistics
PX Logical Circuitry

Logical Circuitry
use CIRCUITS
LOGIC

Logical Product
use BOOLEAN ALGEBRA

Logical Sum
use BOOLEAN ALGEBRA

Long Term
use TIME

M

Machine
use EQUIPMENT

Machine Languages
  use ARTIFICIAL LANGUAGES

MACHINE READABLE DATA
  X Mark Sensing
    Text (Machine-Readable)
  BT Data
  PX Marc (Machine Readable Cataloging)

Machine Translation
  use LANGUAGE TRANSLATION
    COMPUTER APPLICATIONS

Magnetic (Card, Tape, Disc) Typewriters
  use INPUT-OUTPUT EQUIPMENT

MAGNETIC DISCS
  X Discs (Disks)
    Floppy Discs
  BT Storage Media
  RT Magnetic Tape

MAGNETIC TAPE
  X Computer Tape
  BT Tape
  PX Magnetic Discs

Mailgrams
  use CORRESPONDENCE (DOCUMENTS)

Main Entry
  use INDEX TERMS (FORMAT AND INTERRELATIONS)

MAINTENANCE
    Note: Used For Maintenance Of Both Human And Machine Systems.
  BT Uses (Of Resources)
  NT Preservation
    Prevention
    Release
    Restoration
    Retirement
  RT Dysfunctions
  PX File Maintenance

Male
  use MEN

MAN-MACHINE INTERACTION
  X   Adaptive Computer Systems
       Interactive Computer Use (Man-Machine)
  BT   Interaction
  RT   Time-Sharing
  PX   Automated Teaching
       Computer Assisted Indexing
       Computer Assisted Medical Diagnosis
       Learning (Computer-Assisted)

Management
  use ADMINISTRATION

MANAGEMENT INFORMATION SYSTEMS
  X   Business Information Systems
       MIS (Management Information Systems)
  BT   Storage And Retrieval Systems

Manipulative Indexing
  use SEARCHING
       TECHNIQUES

Manpower
  use HUMAN RESOURCES

MANUAL TECHNIQUES
       Note: Used In Conjunction With Terms
             Representing Specific Processes
             or Services.
  BT   Techniques

Manuals
  use BOOKS

MANUFACTURING AND SALES ORGANIZATIONS
  X   Business Organizations
       Companies
       Corporations
  BT   Organizations
  PX   IIA (Information Industry Association)
       Industrial Information Systems
       Information Industry
       Information Industry Association (IIA)
       Publisher

Manuscript
  use RECORDS

MANUSCRIPTS
  BT   Records

Many-To-One Communications
  use COMMUNICATION
       INTERRELATIONS

Maps (Algebraic)
  use ALGEBRAS

Maps (Geographical)
  use GEOGRAPHIC AREAS
       GRAPHIC RECORDS

MARC (Machine Readable Cataloging)
  use MACHINE READABLE DATA
       BIBLIOGRAPHY (PROCESS)

Mark Sensing
  use MACHINE READABLE DATA

MARKETING
  X   Sales
  BT   Finance
  NT   Advertising
       Exhibiting

Masking
  use INTERACTION

Mass Media
  use TELEVISION
       NEWS PUBLICATIONS

Mass Storage
  use STORAGE MEDIA
       QUANTITY

Matching
  use COMPARING

MATERIEL
  BT   Resources
  NT   Personal Property
       Physical Properties
       Substances

Mathematical Models

use MODELING
MATHEMATICS

MATHEMATICS
X Analysis, Mathematical
BT Knowledge Areas
NT Algebras
Geometry
Logic
Statistics
RT Coding Systems
Cybernetics
Evaluation Techniques
Modeling
Numeric Processing
Symbolic Logic
PX American Mathematical Society (AMS)
AMS (American Mathematical Society)
Mathematical Models

Matrices
use ARRANGEMENT

McBee Cards
use EDGE-NOTCHED CARDS

Meaning
use SEMANTICS

MEASURES (MEASURING TECHNIQUES)
X Tests
BT Evaluation
NT Capacity
Effectiveness
Efficiency
Feasibility
Frequency
Incidence
Precision
Productivity
Quality
Quantity
Recall/Precision (Measures)
Reliability
Satisfactoriness
Size
Speed
Strength
Time

Volume
Work (Energy Output)
RT Benefit
Costs
Operations Research
Research
Space (Resource)
Techniques
PX IQ (Intelligence Quotient)

Measuring
use EVALUATION

Mechanized Information Processing
use DATA PROCESSING

Mechanized Literature Searching
use SEARCHING
COMPUTER APPLICATIONS

Mechanized Retrieval
use STORAGE AND RETRIEVAL PROCESSES

Medical Care
use HEALTH CARE

Medical Data
use BIOMEDICAL SCIENCES
DATA

Medical Education
use BIOMEDICAL SCIENCES
INSTRUCTING

Medical Journals
use BIOMEDICAL SCIENCES
PERIODICALS

Medical Library Association (MLA)
use BIOMEDICAL SCIENCES
INFORMATION SCIENCE PROFESSIONAL ORGANIZATIONS

MEDICAL RECORDS
Note: Used In The General Sense; Not Restricted To Patient Records.
X Data, Medical
BT Records
RT Documents

PX Automated Medical History Systems
Computer Assisted Medical Record Systems

Medical Sciences
use BIOMEDICAL SCIENCES

MEDLARS Project (Medical Literature Analysis And Retrieval System)
use BIOMEDICAL SCIENCES
STORAGE AND RETRIEVAL SYSTEMS

Meeting Arrangements
use MEETING ORGANIZATION

MEETING DOCUMENTS
X Agenda (Document)
Attendance Lists
Meeting Planning Manuals
Meetings (Proceedings Or Program Documents)
Minutes Of Meetings
Proceedings (Of Meetings)
Programs Of Meetings (Documents)
BT Documents
RT Meeting Organization
Meetings (Communicating Via Meetings)
Name And Address Lists
Records
Speeches (Documents)

MEETING ORGANIZATION
X Meeting Arrangements
Meeting Planning
Organizing Meetings
Programming (Of Meetings)
BT Interrelations
RT Meeting Documents
Meetings (Communicating Via Meetings)
Scheduling

Meeting Planning
use MEETING ORGANIZATION

Meeting Planning Manuals
use MEETING DOCUMENTS

MEETINGS (COMMUNICATING VIA MEETINGS)
X Clinic (Workshop)
Conferences

Congresses
Institutes (Communicating Via Meetings)
Seminars
Symposia
Workshop (Meeting)
BT Oral Communication
RT Meeting Documents
Meeting Organization
Professional Organizations

Meetings (Proceedings Or Program Documents)
use MEETING DOCUMENTS

Membership (In A Group)
use GROUPS
INTERPERSONAL RELATIONS

Membership Lists
use NAME AND ADDRESS LISTS

MEMORY
Note: Used For Both Human And Electronic Memories.
X Computer Memory
Core Memory
Recall
BT Resources
RT Cognition
Intellect (Intelligence)
Knowledge (State Of Being Informed)
Neurology
Recognition
Storage Media
PX Associative Memory
Content-Addressable Memories
Human Memory
Self-Organizing Systems

MEN
X Male
BT Human Beings
PX Sex (Differences In Employment Or Compensation)
Sex (Differences In Performance)

Mental Disability
use HANDICAPPED PERSONS

Mental Health Care
  use HEALTH CARE

Mental Hospitals
  use HOSPITALS

MERGING
  X Amalgamate
    Cumulation
    Unification
  BT Interrelations
  RT Collating
    Electromechanical Data Processing Equipment
    Selection
    Sorting
  PX Cumulated Indexes

Message (Content Of A Communication)
  use CONCEPTS

Metabolism Of Information
  use INFORMATION FLOW

Metalanguage
  use LANGUAGES

METALLURGY
  BT Physical Sciences
  PX American Society For Metals (ASM)
    ASM (American Society For Metals)

Methods
  use TECHNIQUES

METHODS RESEARCH
    Note: Used For Development Of New Methods For Use in Conducting Research.
  BT Research

Metropolitan
  use CITIES

Mexican Americans
  use ETHNIC GROUPS

MEXICO
  BT Geographic Areas

Microcopies
  use REPRODUCTIONS
      MICROFORM DOCUMENTS

Microfiche (Medium)
  use STORAGE MEDIA
      ARRANGEMENT
      FILM (PHOTOGRAPHIC)

MICROFICHE DOCUMENTS
  X Fiche
  BT Microform Documents

Microfilm (Medium)
  use FILM (PHOTOGRAPHIC)

MICROFILM DOCUMENTS
  BT Microform Documents

Microfilm Readers (Equipment)
  use FILM (PHOTOGRAPHIC)
      READING EQUIPMENT

Microfilm Systems
  use FILM (PHOTOGRAPHIC)
      APPLICATIONS

MICROFORM DOCUMENTS
      Note: Use Only When More Specific Terms
            Will Not Serve Instead.
  BT Audiovisual Software (Documents)
  NT Microfiche Documents
      Microfilm Documents
      Microprint Documents
  RT Storage Media
  PX CIM (Computer Input Microforms)
      COM (Computer Output Microforms)
      Computer Input Microforms (CIM)
      Computer Output Microforms (COM)
      Microcopies
      Micropublishing
      Microreproduction

Microphotography
  use PHOTOGRAPHY

MICROPRINT DOCUMENTS
  BT Microform Documents

PX Microprinting (Publication Process)

Microprinting (Publication Process)
use PRINTING
MICROPRINT DOCUMENTS

MICROPROCESSORS
X Intelligent Computer Terminals
BT Electronic Digital Computers

Micropublishing
use PUBLISHING
MICROFORM DOCUMENTS

Microreproduction
use REPRODUCING (OF DOCUMENTS)
MICROFORM DOCUMENTS

Microthesaurus
use THESAURI

MICROWAVE TECHNIQUES
BT Communication Techniques
RT Telecommunication

Middle Class
use SOCIOECONOMIC GROUPS

MIGRANTS
X Itinerants
BT Roles (Social)

Milieu
use CONDITIONS

MILITARY SCIENCE
BT Behavioral Sciences
RT Defense Documentation Center (DDC)
Department Of Defense
United States Air Force
United States Army
United States Navy

Mind
use INTELLECT (INTELLIGENCE)

MINI-COMPUTERS
BT Electronic Digital Computers

Minority Groups
  use ETHNIC GROUPS
      SOCIOECONOMIC GROUPS

Minutes Of Meetings
  use MEETING DOCUMENTS

MIS (Management Information Systems)
  use MANAGEMENT INFORMATION SYSTEMS

Misdemeanors
  use BEHAVIOR

MLA (Medical Library Association)
  use BIOMEDICAL SCIENCES
      INFORMATION SCIENCE PROFESSIONAL
        ORGANIZATIONS

MODELING
  X Simulation
  BT Creativity
  RT Algebras
      Mathematics
      Techniques
  PX Computer Modeling
      Computer Simulation
      Mathematical Models

Modify
  use CHANGE

Money
  use FINANCE

MONITORING
  BT Surveying
  RT Control Functions
  PX Monitors (Equipment)
      Monitors (Persons)

Monitors (Equipment)
  use EQUIPMENT
      MONITORING

Monitors (Persons)
  use PERSONNEL
      MONITORING

Morale
  use ATTITUDES

Morals
  use VALUES

Morpheme
  use SYNTAX

MOTION PICTURES
  BT Audiovisual Software (Documents)

MOTIVATION
  X Incentives
  BT Goals
  NT Interests
  RT Attitudes
     Compensation
     Values

MOTOR VEHICLES
  X Trucks
  BT Equipment
  RT Transportation

Multidisciplinary Information
  use KNOWLEDGE AREAS
      INFORMATION

Multilingual
  use LANGUAGES

MULTIMEDIA TECHNIQUES
  BT Communication Techniques
  RT Audiovisual Software (Documents)
     Instructing
     Learning
     Oral Communication
     Speech
     Storage Media
     Telecommunication
     Visual Communication
  PX Learning Resource Centers

Multiplexing
  use TIME-SHARING

Multiprocessing
  use DATA PROCESSING

TECHNIQUES

MUMPS (PROGRAMMING LANGUAGE)
BT Artificial Languages

Municipalities
use CITIES

MUSEUMS
BT Input
RT Buildings
Libraries

MUSIC (ART FORM)
BT Fine Arts
RT Dancing
Oral Communication
Sound

MUSIC (DOCUMENTS)
BT Audiovisual Software (Documents)

N

NAME AND ADDRESS LISTS
X Address Lists
Membership Lists
Registration Lists
BT Documents
RT Meeting Documents

NASA (National Aeronautic And Space Administration)
use NATIONAL AERONAUTIC AND SPACE ADMINISTRATION (NASA)

NATIONAL (NATIONWIDE)
BT United States Of America
RT Geographic Areas
Regional
United States Government

National Academy Of Sciences
use SCIENCE AND TECHNOLOGY SERVICE ORGANIZATIONS

NATIONAL AERONAUTIC AND SPACE ADMINISTRATION (NASA)
  X NASA (National Aeronautic And Space Administration)
  BT United States Government

NATIONAL AGRICULTURAL LIBRARY
  BT Libraries

National Auxiliary Publications Service
  use FILM (PHOTOGRAPHIC)
  PUBLISHING
  SERVICE ORGANIZATIONS

NATIONAL BUREAU OF STANDARDS
  BT United States Government

National Federation Of Abstracting And Indexing Services
  use ABSTRACTING
  INDEXING
  INFORMATION SCIENCE SERVICE ORGANIZATIONS

National Government (U.S.)
  use UNITED STATES GOVERNMENT

NATIONAL LIBRARY OF MEDICINE
  BT Libraries

NATIONAL REFERRAL CENTER
  BT United States Government

NATIONAL SCIENCE FOUNDATION
  X NSF
  BT United States Government
  PX Office Of Science And Technology (NSF)
  Office Of Scientific Information Services (OSIS)
  Osis (Office Of Scientific Information Services)

National Technical Information Service (NTIS)
  use INFORMATION SCIENCE SERVICE ORGANIZATIONS
  UNITED STATES GOVERNMENT
  SCIENCE AND TECHNOLOGY

Native Speakers
  use SPEECH
  NATURAL LANGUAGES

GEOGRAPHIC AREAS

Natural Language Indexes
  use INDEXES
      NATURAL LANGUAGES

NATURAL LANGUAGES
  X   Foreign Language
      Idioms
  BT Languages
  NT Arabic Language
      Chinese Language
      English Language
      French Language
      German Language
      Japanese Language
      Russian Language
      Spanish Language
  PX Free-Text Indexing
      Language Training
      Native Speakers
      Natural Language Indexes
      Search Languages (Natural Languages)
      Vernacular

NEED
  X   Requirements (Need)
  BT Input
  NT Information Need
  PX User Needs

Neighborhood
  use COMMUNITIES

NERVOUS SYSTEM
  X   Neural Networks
      Nodes (Neural)
  BT Neurology
  NT Senses
  RT Speech

Networks, Time-Sharing
  use TIME-SHARING SERVICES

Neural Networks
  use NERVOUS SYSTEM

NEUROLOGY
  BT Biomedical Sciences

NT Nervous System
RT Human Beings
Intellect (Intelligence)
Memory

NEWS PUBLICATIONS
X Announcement Media
Bulletins
Newspapers
BT Documents
RT Advertising Literature
Disseminating
Lay Publications
Publications (Personal Or Organizational)
Television
PX Mass Media

Newspapers
use NEWS PUBLICATIONS

NIGHT
BT Time

Nodes (Neural)
use NERVOUS SYSTEM

NOISE (IN COMMUNICATION SYSTEMS)
BT Dysfunctions

Noise (Sound)
use SOUND

Nomenclature
use CLASSIFICATION SYSTEMS

Nominating
use SELECTION

Nonbook Materials
use AUDIOVISUAL SOFTWARE (DOCUMENTS)

Nonprofit Organizations
use ORGANIZATIONS
FINANCE

NORMAL FUNCTIONS
BT Output
NT Responses
PX Library Operations

NORWAY
  BT   Europe

Notation Systems
  use   CODING SYSTEMS

NSF
  use   NATIONAL SCIENCE FOUNDATION

NTIS (National Technical Information Service)
  use   INFORMATION SCIENCE SERVICE ORGANIZATIONS
        UNITED STATES GOVERNMENT
        SCIENCE AND TECHNOLOGY

Null Hypothesis
  use   THEORY

NUMBERING SYSTEMS
  X   Binary Numbering
       Octal Numbering
  BT   Symbol Sets
  RT   Advertising Literature

Numeric Coding
  use   CODING

NUMERIC PROCESSING
  X   Computation
  BT   Data Processing
  NT   Calculations
  RT   Data Processing Services
       Finance
       Graphic Records
       Mathematics
       Statistics
       Word Processing

Nursing
  use   HEALTH CARE

Nutrition
  use   BIOMEDICAL SCIENCES

O

Obituaries
  use   BIOGRAPHICAL DOCUMENTS

Objectives
  use GOALS

OBSERVATIONAL RESEARCH
  BT Research

Obstacles
  use BARRIERS

Occupations
  use EMPLOYMENT

Octal Numbering
  use NUMBERING SYSTEMS

OFFICE OF EDUCATION
  BT United States Government

Office Of Science And Technology (NSF)
  use NATIONAL SCIENCE FOUNDATION
      SERVICE ORGANIZATIONS

Office Of Scientific Information Services (OSIS)
  use NATIONAL SCIENCE FOUNDATION
      INFORMATION SCIENCE SERVICE ORGANIZATIONS

OFFLINE DATA PROCESSING
  BT Data Processing

Offset Printing (Process)
  use PRINTING

Offset-Printing Equipment
  use PRINTING EQUIPMENT

One-To-Many Communication
  use COMMUNICATION
      INTERRELATIONS

ONLINE DATA PROCESSING
  BT Data Processing
  RT Time-Sharing
     Time-Sharing Services

OPERATION (OF EQUIPMENT)
  BT Applications
  PX Operators

OPERATIONS RESEARCH
BT Research
RT Control Functions
Evaluation Techniques
Measures (Measuring Techniques)
PX Library Research (Evaluation Of Library Functions)

Operators
use OPERATION (OF EQUIPMENT)
PERSONNEL

Opinions
use ATTITUDES

OPPORTUNITY
BT Interrelations
RT Human Resources
Resources

Optical Aids
use EQUIPMENT
VISION

Optical Character Recognition
use OPTICAL SCANNERS

Optical Coincidence Cards
use OPTICAL COINCIDENCE TECHNIQUES

OPTICAL COINCIDENCE TECHNIQUES
X Batten System
Filmorex System
Optical Coincidence Cards
Peek-A-Boo Cards
BT Data Processing Techniques
RT Cards
Comparing
Electromechanical Data Processing Equipment
Punched Cards

Optical Display
use VISUAL COMMUNICATION

OPTICAL SCANNERS
X Character Recognition Equipment
Optical Character Recognition
BT Reading Equipment

ORAL COMMUNICATION
X Reporting (Oral)
BT Communication
NT Meetings (Communicating Via Meetings)
Singing
Speech
RT Audiovisual Equipment
Audiovisual Software (Documents)
Meetings
Multimedia Techniques
Music (Art Form)
Sound
Speeches (Documents)
Telephone
Television
Visual Communication
PX Conference Calls (Telephone)
Oral Histories

Oral Histories
use ORAL COMMUNICATION
DOCUMENTS

Order (Purchase)
use ACQUIRING

Ordering (Arranging)
use INTERRELATIONS

Organic Chemistry
use CHEMISTRY

Organization
use INTERRELATIONS

ORGANIZATIONS
X Agencies
Associations
Centers
Institutions
Location (Organizational Affiliation)
Societies
BT Input
NT Academic Organizations
Governments
Manufacturing And Sales Organizations
Philanthropic Organizations
Professional Organizations

Service Organizations
RT Bylaws
International (Geography)
PX Elective Offices
Internal Reports
Labor Unions
Nonprofit Organizations
Organized Labor

Organized Labor
use EMPLOYMENT ORGANIZATIONS

Organizing
use INTERRELATIONS

Organizing Meetings
use MEETING ORGANIZATION

Orientation
use RELATIONS

OSIS (Office Of Scientific Information Services)
use NATIONAL SCIENCE FOUNDATION
INFORMATION SCIENCE SERVICE ORGANIZATIONS

Outcome
use OUTPUT

Outlook
use ATTITUDES

OUTPUT
X Outcome
BT Information Science
NT Dysfunctions
Normal Functions

Output Data
use OUTPUT RECORDS

OUTPUT RECORDS
X Hardcopy
Output Data
Printouts
BT Records
NT Samples
RT Input Records

Overtime
  use WAGES

Ownership
  use CONTROL FUNCTIONS

P

PAINTINGS
  BT Graphic Records
  RT Fine Arts
     Photographs

PAKISTAN
  BT Asia

PAMPHLETS
  X Brochures
  BT Documents

Paper
  use STORAGE MEDIA

PAPER TAPE
  X Chadless Tape
    Perforated Tape
  BT Tape

Papers (Records)
  use RECORDS

Parameters
  use CONDITIONS

PARAPROFESSIONALS
  BT Personnel
  RT Volunteers
  PX Library Assistants
     Aides

Parking Facilities
  use FACILITIES

Parsing
  use SYNTAX

Part-Time Personnel

use PERSONNEL

Participation
use INTERACTION

Passivity
use BEHAVIOR

PATENTS
BT Documents
PX Chemical Patents

PATIENTS
BT Roles (Social)

PATTERN RECOGNITION
BT Recognition

Patterns
use INTERRELATIONS

Payment
use FINANCE

PB Reports (Publication Board Reports)
use RECORDS
SCIENCE AND TECHNOLOGY

Peace Corps
use INSTRUCTING
INTERNATIONAL (GEOGRAPHY)
UNITED STATES GOVERNMENT

Peek-A-Boo Cards
use OPTICAL COINCIDENCE TECHNIQUES

PEERS
BT Roles (Social)
RT Personnel
PX Refereeing

Pensions
use WAGES

Perception
use COGNITION

Perforated Tape
use PAPER TAPE

PERFORMANCE
X Practice
Work Performance
BT Behavior
PX Index Performance
Job Performance
Sex (Differences In Performance)

PERIODICALS
X Serials
BT Documents
RT Publications (Personal Or Organizational)
PX American Documentation (Journal)
Coden
Medical Journals
Reprints (Or Periodical Articles)

Permanent
use TIME

PERMUTATION
BT Interrelations
RT KWIC Indexing
PX Chain Indexing

Person-To-Machine Communication
use MAN-MACHINE INTERACTION

PERSON-TO-PERSON COMMUNICATION
BT Communication

Personal Adjustment
use BEHAVIOR

Personal Indexes (Of Scientists)
use INDEXES
SCIENTISTS

PERSONAL PROPERTY
X Property (Personal)
Real Estate
BT Materiel

Personal Rights
use RIGHTS

PERSONALITY
X Emotions
BT Human Resources

NT Attitudes

PERSONNEL
X Assistants
Attendants
Employees
Part-Time Personnel
Staff (Employees)
System Operators
BT Roles (Social)
NT Information Science Personnel
Paraprofessionals
Scientists
RT Behavior
Consulting Services
Employment
Human Relations
Input
Peers
Rules
PX Administrators
Aides
Attorneys
Consultants
Coordinators
Designers
Executives
Experts
Faculty
Health Professionals
Investigators
Lawyers
Librarians
Monitors (Persons)
Operators
Police
Programmers
Readers (Proof-Readers)
Research Investigators
Students
Trainees
Work Study Program

PERT (PROGRAM EVALUATION REVIEW TECHNIQUES)
X Program Evaluation Review Techniques (PERT)
BT Data Analysis

Petroleum Science And Technology

use SCIENCE AND TECHNOLOGY

Pharmaceutical Sciences
use BIOMEDICAL SCIENCES

Pharmacology
use BIOMEDICAL SCIENCES

PHILANTHROPIC ORGANIZATIONS
X Ford Foundation
BT Organizations

PHILOSOPHY
BT Knowledge Areas

Phonation
use SPEECH

Phonemes
use SYNTAX

Phonetic Indexing
use INDEXING
SOUND

Phonetics
use SPEECH

Phonograph
use AUDIOVISUAL EQUIPMENT

PHOTOCOMPOSITION
BT Composing

Photocopies
use REPRODUCTIONS

PHOTOCOPYING
BT Reproducing (Of Documents)
RT Electrostatic Processes
Reproducing Equipment
PX Fair Use Copying

Photocopying Equipment
use REPRODUCING EQUIPMENT

PHOTOGRAPHS
X Illustrations (Photographic)
Pictures

BT Graphic Records
NT Holograms
RT Paintings
Reproducing (Of Documents)
PX Photointerpretation

PHOTOGRAPHY
X Filming
Microphotography
BT Techniques
NT Holography
RT Film (Photographic)
Printing
PX Cameras

Photointerpretation
use INTERPRETING
PHOTOGRAPHS

PHYSICAL PROPERTIES
BT Materiel
NT Temperature

PHYSICAL SCIENCES
X Crystallography
BT Science And Technology
NT Aerospace Sciences
Atomic And Molecular Physics
Cybernetics
Electronics
Engineering
Geology
Metallurgy
RT Biological Sciences
Chemistry
Light
Sound
PX AIP (American Institute Of Physics)
American Institute Of Physics (AIP)
Physicists

Physicians
use BIOMEDICAL SCIENCES
SCIENTISTS

Physicists
use PHYSICAL SCIENCES
SCIENTISTS

Picturephones
  use TELEPHONE
      TELEVISION

Pictures
  use PHOTOGRAPHS

PL/1 (PROGRAMMING LANGUAGE)
  BT Artificial Languages

PLANNING
  X Agenda
    Allocation
    Device (Plan)
  BT Decision-Making
  NT Designing
  RT Creativity
    Feasibility
    Finance
  PX City Planning
    PPBS (Program Planning And Budgeting System)
    Program (Of Activities)
    Program Planning And Budgeting System (PPBS)

PLATO SYSTEM
  BT Computer Assisted Instructing (CAI)

Playback Systems
  use AUDIOVISUAL EQUIPMENT

Plotter (Equipment)
  use GRAPHICS PRINTING EQUIPMENT

POLAND
  BT Europe

Police
  use LAW
      PERSONNEL

POLICIES
  BT Rules

Policing
  use CONTROL FUNCTIONS

POLITICAL SCIENCE
  BT   Behavioral Sciences

Poor
  use   SOCIOECONOMIC GROUPS

POPULATION
  X   Citizens
       Controls (Research)
       Residents
       Samples (Statistical)
       Subjects (Research)
  BT   Resources
  NT   Animals
       Human Beings
  RT   Roles (Social)
  PX   British (People)
       Census

POSTGRADUATE EDUCATION
  X   Continuing Education
       Internship
  BT   Education
  RT   Courses (Educational)
       Curricula
       Instructing
       Learning
  PX   Fellowships

Potential
  use   RESOURCES

PPBS (Program Planning And Budgeting System)
  use   PLANNING
        BUDGETING

Practice
  use   PERFORMANCE

PRECISION
  BT   Measures (Measuring Techniques)
  RT   Recall/Precision (Measures)

PREDICTING
  X   Prognosis
       Projection
  BT   Statistics
  RT   Forecasts (Documents)
       Time

Preference
  use INTERESTS

Prejudice
  use BIAS

PRESERVATION
  BT Maintenance
  RT Recording
     Storage Media

PREVENTION
  X Impede
  BT Maintenance

Price/Value
  use COSTS
      EFFECTIVENESS

PRINTING
  X Offset Printing (Process)
  BT Techniques
  RT Composing
     Photography
     Publishing
     Recording
     Reproducing (Of Documents)
     Symbol Sets
  PX Automated Typesetting
     Microprinting (Publication Process)

PRINTING EQUIPMENT
  X Composition Equipment
    Line Printers
    Offset-printing Equipment
    Typewriters
  BT Input-Output Equipment
  NT Graphics Printing Equipment
  RT Composing
  PX Type Fonts

PRINTING SERVICES
  BT Delivery Of Services
  RT Audiovisual Production Services
     Copying Services
     Publishing Services

Printouts

use OUTPUT RECORDS

PRIORITIES
BT Values
RT Scheduling
Space (Resource)
Time
Time-Sharing

PRISONS
BT Buildings

PRIVACY
BT Independence

Private
use RELATIONS

PROBABILITY
BT Statistics
NT Randomness
Weighting
RT Relations
Risk

Problem-Solving
use FEEDBACK AND CONTROL

Procedures
use TECHNIQUES

Proceedings (Of Meetings)
use MEETING DOCUMENTS

PROCESSES (PROCESSING)
BT Information Science
NT Communication
Data Processing
Reproducing (Of Documents)
Transportation
RT Conditions

Procurement (Of Documents)
use ACQUIRING

PRODUCTIVITY
BT Measures (Measuring Techniques)
RT Capacity

Professional Behavior
  use HUMAN RELATIONS
      ETHICS

Professional Literature
  use DOCUMENTS

PROFESSIONAL ORGANIZATIONS
  BT Organizations
  NT Information Science Professional
        Organizations
  RT Consulting Services
      Meetings (Communicating Via Meetings)
  PX ABA (American Bar Association)
      AIBS (American Institute Of Biological
        Sciences)
      AIP (American Institute Of Physics)
      AMA (American Management Association)
      AMA (American Medical Association)
      American Bar Association (ABA)
      American Institute Of Biological Sciences
        (AIBS)
      American Institute Of Physics (AIP)
      American Management Association (AMA)
      American Mathematical Society (AMS)
      American Medical Association (AMA)
      American Psychological Association (APA)
      American Society For Metals (ASM)
      AMS (American Mathematical Society)
      APA (American Psychological Association)
      ASM (American Society For Metals)
      IEEE (Institute Of Electrical And
        Electronic Engineers)
      Institute Of Electrical And Electronic
        Engineers (IEEE)
      International Federation For Documentation
        (FID)

Professional Training
  use UNIVERSITY AND COLLEGE EDUCATION

Professions
  use EMPLOYMENT

Professors
  use INSTRUCTING
      INFORMATION SCIENCE PERSONNEL

Profiles (Of Interests)
  use INTERESTS

Prognosis
  use PREDICTING

Program (Of Activities)
  use PLANNING
    APPLICATIONS

Program Evaluation Review Techniques (PERT)
  use PERT (PROGRAM EVALUATION REVIEW TECHNIQUES)

Program Planning And Budgeting System (PPBS)
  use PLANNING
    BUDGETING

Programmed Instruction
  use PROGRAMS
    INSTRUCTING

Programmed Learning
  use PROGRAMS
    LEARNING

Programmers
  use PROGRAMMING
    PERSONNEL

PROGRAMMING
  X Batch Programming
    Computer Programming
  BT Data Processing Techniques
  PX Programmers

Programming (Of Meetings)
  use MEETING ORGANIZATION

Programming Languages
  use ARTIFICIAL LANGUAGES

PROGRAMS
    Note: Used For Communicating Via Formalized Procedures.
  X Algorithms
    Computer Software
    Instructions (Algorithmic)
    Software (Computer Programs)
  BT Communication Techniques

RT Curricula
Input Records
Instructing
Recognition
Time-Sharing
PX Assembling (Computer Programs)
Compiling (Computer Programs)
Indexing Instructions (To Data Processing Machines)
Programmed Instruction
Programmed Learning
Report Generators
Syllabus
Transformational Grammars

Programs Of Meetings (Documents)
use MEETING DOCUMENTS

Projection
use PREDICTING

Projects
use APPLICATIONS

Proofreading
use ERROR DETECTION AND CORRECTION

Property (Personal)
use PERSONAL PROPERTY

PROPOSALS
X Bidding
Request For Bid (RFP)
RFP (Request For Bid)
BT Documents
PX Contract Proposals
Grant Proposals

Psycholinguistics
use LINGUISTICS
PSYCHOLOGY

PSYCHOLOGY
BT Behavioral Sciences
NT Behavior
PX Psycholinguistics

Public Health
use BIOMEDICAL SCIENCES

PUBLIC HEALTH SERVICE
  X   FDA (United States Food And Drug Administration)
       United States Food And Drug Administration (FDA)
  BT   Health, Education And Welfare Department (HEW)

Public Libraries
  use LIBRARIES

Public Relations
  use HUMAN RELATIONS

PUBLICATIONS (PERSONAL OR ORGANIZATIONAL)
       Note: Used Only When Specific Types Of Documents Cannot Be Named, Such As Other Terms That Are Hierarchically Narrower Than "Documents."
  X   House Organs
  BT   Documents
  RT   Abstracting
       Composing
       Lay Publications
       News Publications
       Periodicals
       Records

Publicity
  use HUMAN RELATIONS

Publisher
  use PUBLISHING
       MANUFACTURING AND SALES ORGANIZATIONS

PUBLISHING
  BT   Communication
  RT   Authors
       Composing
       Disseminating
       Printing
       Techniques
       Writing (Creative)
  PX   APS (Auxiliary Publication Service)
       Auxiliary Publication Service (APS)
       Micropublishing
       National Auxiliary Publications Service
       Publisher

PUBLISHING SERVICES
  BT   Delivery Of Services
  RT   Audiovisual Production Services
       Printing Services

Punched Card Systems
  use APPLICATIONS
       PUNCHED CARDS

PUNCHED CARDS
  X   Hollerith Cards
       IBM Cards
       Tabulating Cards
  BT   Cards
  NT   Edge-Notched Cards
  RT   Electromechanical Data Processing Equipment
       Input-Output Equipment
       Optical Coincidence Techniques
  PX   Punched Card Systems

Punching Equipment
  use INPUT-OUTPUT EQUIPMENT

Purchase Cost
  use COSTS

Purchasing
  use ACQUIRING

Q

QUALITY
  BT   Measures (Measuring Techniques)
  PX   Quality Control

Quality Control
  use CONTROL FUNCTIONS
       QUALITY

QUANTITY
  BT   Measures (Measuring Techniques)
  PX   Mass Storage
       Quantity Control

Quantity Control
  use CONTROL FUNCTIONS
       QUANTITY

Queries
  use INQUIRIES

Query Formulation
  use INQUIRIES
      INTERRELATIONS

Query Languages (Conrolled)
  use THESAURI

Question-Answering Systems (Automated)
  use RESPONSES (TO INQUIRIES)
      CONCEPTS
      COMPUTER APPLICATIONS

QUESTIONNAIRE TECHNIQUES
  BT Data Gathering
  RT Interviewing
     Surveying

QUESTIONNAIRES
  BT Instruments
  RT Documents
     Surveying

Questions
  use INQUIRIES

QUEUING THEORY
  BT Theory

R

Race
  use ETHNIC GROUPS

RADIO
  BT Telecommunication
  PX Radiograms

Radiograms
  use CORRESPONDENCE (DOCUMENTS)
      RADIO

Random Access
  use ACCESSING
      STORAGE MEDIA

Random Access Storage
  use STORAGE MEDIA
    INTERRELATIONS

RANDOM ALARM TECHNIQUES
  BT Data Gathering
  RT Surveying

Random Coding
  use CODING
    RANDOMNESS

RANDOMNESS
  BT Probability
  RT Interrelations
    Sampling
  PX Random Coding
    Superimposed Coding

RANKING
  X Scaling
    Scoring
  BT Interrelations

Readability
  use READING

Reader-Printer
  use INPUT-OUTPUT EQUIPMENT

Readers (Library Users)
  use USERS (OF SERVICES)

Readers (Proof-Readers)
  use ERROR DETECTION AND CORRECTION
    PERSONNEL

READINESS
  BT Interrelations
  RT Availability

READING
  X Bibliotherapy
    Readability
  BT Visual Communication
  RT Input Process
  PX Automated Reading
    Illiteracy
    Literacy

READING EQUIPMENT
X Viewers (Equipment)
BT Input-Output Equipment
NT Optical Scanners
PX Microfilm Readers (Equipment)
Tape Reading Equipment

Real Estate
use PERSONAL PROPERTY

Real-Time
use TIME
DATA PROCESSING

REASONING
BT Intellect (Intelligence)

Recall
use MEMORY

RECALL/PRECISION (MEASURES)
BT Measures (Measuring Techniques)
RT Precision

RECEIVERS (EQUIPMENT)
BT Input-Output Equipment
PX Transceivers (Tranmsitters/Receivers) (Equipment)

Recipients (Receivers) (Human)
use CLIENTELE

Reciprocal
use RELATIONS

RECOGNITION
BT Intellect (Intelligence)
NT Pattern Recognition
RT Criteria
Learning
Memory
Programs
PX Character Recognition
Content Recognition
Discrimination (Electronic Object Detection)

Recommendations
use SPECIFICATIONS

Recorded Books
  use SOUND RECORDINGS
      BOOKS

RECORDING
  BT Communication
  NT Writing
  RT Preservation
      Printing
  PX Electronic Videorecording (EVR)
      EVR (Electronic Videorecording)

Recordings (Sound)
  use SOUND RECORDINGS

RECORDS
  BT Documents
  NT Input Records
      Manuscripts
      Medical Records
      Output Records
      Research Results
      Surveys (Records)
  RT Correspondence (Documents)
      Graphic Records
  PX Report Generators

RECREATION
  BT Behavior

Recruiting
  use EMPLOYMENT

REDUNDANCY
  BT Interaction

Reeducation
  use INSTRUCTING

Refereeing
  use CONTROL FUNCTIONS
      PEERS

Reference Questions
  use INQUIRIES

Reference Services
  use RESPONSES (TO INQUIRIES)

References (For Employment)
  use EMPLOYMENT
      CRITERIA

References (To Documents)
  use CITATIONS

Referral
  use SWITCHING
      CLIENTELE

REGIONAL
      Note: Used To Connote Either Geographic
            Areas Or Regional Activities.
  BT United States Of America
  RT Cities
      Geographic Areas
      International (Geography)
      National (Nationwide)
      Rural
      Suburban
  PX Regional Information Centers
      Regional Medical Libraries
      Regional Service Centers

Regional Information Centers
  use REGIONAL
      INFORMATION SCIENCE SERVICE ORGANIZATIONS

Regional Medical Libraries
  use REGIONAL
      BIOMEDICAL SCIENCES
      LIBRARIES

Regional Service Centers
  use REGIONAL
      SERVICE ORGANIZATIONS

Registration Lists
  use NAME AND ADDRESS LISTS

Registry Systems
  use CLASSIFICATION SYSTEMS

Regulations
  use RULES

Rehabilitation

use DYSFUNCTIONS
CHANGE

Reinforcement
use CONDITIONING

Rejection
use SELECTION

RELATIONS
X Dependent
Impact
Orientation
Private
Reciprocal
Systems Phenomena
BT Environment
NT Barriers
Human Relations
Independence
International Relations
RT Communication
Interrelations
Probability
Resources

RELEASE
BT Maintenance
RT Independence

Relevance
use EVALUATION

Relevance Studies
use EVALUATION
RESPONSES (TO INQUIRIES)

RELIABILITY
BT Measures (Measuring Techniques)
RT Effectiveness
Efficiency
Feasibility

Remedy
use CHANGE

Remunerate
use COMPENSATION

Rental
  use FINANCE

Report Generators
  use CONSTRUCTING
      PROGRAMS
      RECORDS

Reporting (Oral)
  use ORAL COMMUNICATION

Reporting (Written)
  use RECORDS

Reports
  use RECORDS

REPRESENTATIVES
  BT Roles (Social)

Reprints
  use REPRODUCTIONS

Reprints (Or Periodical Articles)
  use REPRODUCTIONS
      PERIODICALS

REPRODUCING (OF DOCUMENTS)
  X Copying (Of Documents)
    Image-Making (Of Documents)
    Reprography
  BT Processes (Processing)
  NT Electrostatic Processes
     Photocopying
  RT Film (Photographic)
     Holograms
     Photographs
     Printing
     Storage Media
  PX Card Reproduction
     Duplicating (Documents)
     Microreproduction

REPRODUCING EQUIPMENT
  X Photocopying Equipment
  BT Input-Output Equipment
  RT Film (Photographic)
     Photocopying

Reproductions
Storage Media

REPRODUCTIONS
X Copies (Of Documents)
Facsimiles
Images, Document
Photocopies
Reprints
BT Documents
RT Reproducing Equipment
PX Facsimile Transmission
Filmed Documents
Microcopies
Reprints (Or Periodical Articles)

Reprography
use REPRODUCING (OF DOCUMENTS)

Request For Bid (RFP)
use PROPOSALS

Requirements (Need)
use NEED

Requirements (Specifications)
use SPECIFICATIONS

RESEARCH
X Explore
BT Evaluation
NT Archival Research
Experimental Research
Methods Research
Observational Research
Operations Research
RT Measures (Measuring Techniques)
Statistics
Techniques
Theory
Variables
PX Advanced Research Projects Agency (ARPA)
ARPA (Advanced Research Projects Agency)
Arpanet
Hypothesis Testing
ILR (Institute Of Library Research)
Institute Of Library Research (ILR)
Research Investigators

Research Organizations
Science Information Exchange (SIE)
SIE (Science Information Exchange)

Research Investigators
use RESEARCH
PERSONNEL

Research Libraries
use LIBRARIES
ACADEMIC ORGANIZATIONS

Research Organizations
use RESEARCH
SERVICE ORGANIZATIONS

RESEARCH RESULTS
X Findings
BT Records
RT Data
Data Analysis
PX Research Utilization

Research Utilization
use RESEARCH RESULTS
USES (OF RESOURCES)

Reservation Systems
use SCHEDULING
STORAGE AND RETRIEVAL SYSTEMS

Reserve Books
use BOOKS
SCHEDULING

RESIDENTIAL FACILITIES
X Houses
BT Buildings

Residents
use POPULATION

RESISTANCE (HUMAN)
BT Behavior

Resource Management
use RESOURCES
ADMINISTRATION

RESOURCES
  X   Potential
       Sources
  BT   Environment
  NT   Facilities
       Geographic Areas
       Human Resources
       Knowledge Areas
       Languages
       Light
       Materiel
       Memory
       Population
       Sound
       Space (Resource)
  RT   Conditions
       Decision-Making
       Goals
       Opportunity
       Relations
       Roles (Social)
       Rules
  PX   Information Resources
       Library Resources
       Resource Management

Response Time
  use RESPONSES
       TIME

RESPONSES
  BT   Normal Functions
  NT   Delivery Of Services
  PX   Response Time

Responses (Dysfunctional)
  use DYSFUNCTIONS

RESPONSES (TO INQUIRIES)
  X   Information Services (Responses To
            Inquiries)
       Reference Services
       Searches
  BT   Delivery Of Services
  RT   Bibliographies
       Information Need
       Thesauri
  PX   Question-Answering Systems (Automated)
       Relevance Studies

RESPONSIBILITY
X Accountability
Liability
BT Behavior
RT Administration
Control Functions

RESTORATION
BT Maintenance

Resumes (Of Documents)
use ABSTRACTS (OF DOCUMENTS)

Resumes (Of Persons)
use BIOGRAPHICAL DOCUMENTS

RETIREMENT
BT Maintenance
RT Employment
Retirement Plans

Retirement (Of Goods)
use DISPOSAL

RETIREMENT PLANS
BT Compensation
RT Retirement

Retrieval (Process)
use STORAGE AND RETRIEVAL PROCESSES

Retrieval Effectiveness
use RESPONSES (TO INQUIRIES)
EFFECTIVENESS

Retrieval Languages (Controlled)
use THESAURI

Retrospective Searching
use DATA GATHERING

Reversal
use INTERRELATIONS
CHANGE

Reviewing Process
use EVALUATION

REVIEWS
X Annotating
Annual Reviews
Assessing
BT Documents
RT Abstracts (Of Documents)
Criteria
Evaluation
Surveying
PX Annotated Bibliographies
Annual Review Of Information Science And Technology (ARIST)
ARIST (Annual Review Of Information Science And Technology)
Critical Reviews
State-Of-The-Art Studies

Revision
use ERROR DETECTION AND CORRECTION

Rewards
use COMPENSATION

RFP (Request For Bid)
use PROPOSALS

Rhythm
use SEQUENCING

RIGHTS
X Civil Liberties
Legal Rights
Personal Rights
BT Rules
PX ACLU (American Civil Liberties Union)
American Civil Liberties Union (ACLU)

Riots
use DYSFUNCTIONS

RISK
BT Interaction
RT Probability

Roles (Indexing)
use INDEX TERMS (FORMAT AND INTERRELATIONS)

ROLES (SOCIAL)
BT Environment

NT Adults
Applicants
Authors
Children
Clientele
Consumers
Family
Migrants
Patients
Peers
Personnel
Representatives
Volunteers
RT Population
Resources

ROUTING
BT Disseminating
RT Transmission

RULES
X Eligibility
Guides (Rules)
Ineligibility
Regulations
BT Environment
NT Laws
Policies
Rights
RT Law
Personnel
Resources
Rights

RURAL
Note: Used For Either Rural Geography Or Activities Which Are Typically Rural.
BT United States Of America
RT Geographic Areas
Regional

RUSSIA
BT Europe
PX Russian Literature

RUSSIAN LANGUAGE
BT Natural Languages

Russian Literature
  use DOCUMENTS
      RUSSIA

S

Safety
  use SECURITY

Salaries
  use WAGES

Sales
  use MARKETING

SAMPLES
  BT Output Records
  RT Exhibits (Documents)

Samples (Display Objects; Tokens)
  use EXHIBITING

Samples (Statistical)
  use POPULATION

SAMPLING
  BT Statistics
  RT Data Gathering
     Randomness
     Variables

SATISFACTION
  BT Human Relations
  X Job Satisfaction

SATISFACTORINESS
  BT Measures (Measuring Techniques)

Savings
  use FINANCE

Scaling
  use RANKING

SCHEDULING
  BT Sequencing

RT Meeting Organization
Priorities
Time
PX Computer Scheduling
Reservation Systems
Reserve Books

Schematics
use GRAPHIC RECORDS

Scholarships
use AWARDS

Schools
use ACADEMIC ORGANIZATIONS

SCIENCE AND TECHNOLOGY
X Information, Science
Petroleum Science And Technology
Technology
BT Knowledge Areas
NT Behavioral Sciences
Biological Sciences
Chemistry
Physical Sciences
PX American Petroleum Institute (API)
API (American Petroleum Institute)
ICSU (International Council Of Scientific Unions)
Indian National Scientific Documentation Center (INDSDOC)
INSDOC (Indian National Scientific Documentation Center)
International Council Of Scientific Unions (ICSU)
Japan Information Center Of Science And Technology (JICST)
JICST (Japan Information Center Of Science And Technology)
National Academy Of Sciences
National Technical Information Service (NTIS)
NTIS (National Technical Information Service)
PB Reports (Publication Board Reports)
Science Information
Science Information Exchange (SIE)
Scientific Writing

SIE (Science Information Exchange)
Technical Writing

Science Information
use SCIENCE AND TECHNOLOGY
INFORMATION

Science Information Exchange (SIE)
use SCIENCE AND TECHNOLOGY
RESEARCH
INFORMATION SCIENCE SERVICE ORGANIZATIONS

Scientific Writing
use SCIENCE AND TECHNOLOGY
WRITING (CREATIVE)

SCIENTISTS
BT Personnel
PX Biologists
Engineers
Personal Indexes (Of Scientists)
Physicians
Physicists
Social Scientists

Scoring
use RANKING

Screening
use SELECTION

SCULPTURE
BT Fine Arts

SDI
Note: Abbreviation For "Selective Dissemination Of Information."
X Selective Dissemination Of Information (SDI)
BT Disseminating
RT Information Flow

Search Languages (Controlled)
use THESAURI

Search Languages (Natural Languages)
use SEARCHING
NATURAL LANGUAGES

Search Strategy
  use SEARCHING
  INTERRELATIONS
  TECHNIQUES

Searches
  use RESPONSES (TO INQUIRIES)

SEARCHING
  X Bibliographic Search
  Information Searching
  Iterative Search Processes
  Library Research (Literature Searches)
  Literary Research (Searching The Literature)
  BT Storage And Retrieval Processes
  RT Comparing
  Selection
  Storage And Retrieval Systems
  Thesauri
  PX Data Bases (For Automated Storage And Retrieval)
  Information Collecting (Use Of Indexes)
  Legal Research (Searching Legal Literature)
  Manipulative Indexing
  Mechanized Literature Searching
  Search Languages (Natural Languages)
  Search Strategy

Secondary Education
  use HIGH SCHOOL EDUCATION

Secondary Publications (Indexes; Abstracts Publications; Reviews)
  use BIBLIOGRAPHY (PROCESS)

Secondary Services (Storage And Retrieval)
  use INFORMATION SCIENCE SERVICE ORGANIZATIONS

SECURITY
  X Safety
  BT Interaction
  PX File Security

See Also Reference
  use INDEX TERMS (FORMAT AND INTERRELATIONS)

See Reference
  use INDEX TERMS (FORMAT AND INTERRELATIONS)

SELECTION
Nominating
Rejection
Screening
BT Data Gathering
RT Comparing
Criteria
Merging
Searching
Sorting
Storage And Retrieval Processes
Thesauri
PX False Drops (Unwanted Selections)

Selective Dissemination Of Information (SDI)
use SDI

SELF-INSTRUCTING
BT Instructing
RT Libraries

Self-Organizing Systems
use MEMORY
INTERRELATIONS

Semantic Factoring
use FACTOR ANALYSIS

SEMANTICS
X Meaning
Synonyms
BT Linguistics
RT Coding Systems
Concepts

Seminars
use MEETINGS (COMMUNICATING VIA MEETINGS)

Senior Citizens
use AGED

SENSES
BT Nervous System
NT Hearing
Smell
Taste
Touch
Vision

RT Light
Sound

Sentence
use SYNTAX

SEQUENCING
X Rhythm
BT Interrelations
NT Scheduling
RT Sorting

Serials
use PERIODICALS

SERVICE ORGANIZATIONS
X Utilities
BT Organizations
NT Information Science Service Organizations
Voluntary Services
RT Abstracting
Consulting Services
Delivery Of Services
Information
Libraries
PX ACLU (American Civil Liberties Union)
ACS (American Chemical Society)
American Chemical Society (ACS)
American Civil Liberties Union (ACLU)
American National Standards Institute (ANSI)
American Petroleum Institute (API)
American Standards Organization
ANSI (American National Standards Institute)
API (American Petroleum Institute)
Civic Groups
General Services Administration (GSA)
GSA (General Services Administration)
ICSU (International Council Of Scientific Unions)
Information Industry
International Council Of Scientific Unions (ICSU)
International Standards Organization (ISO)
ISO (International Standards Organization)
National Academy Of Sciences
National Auxiliary Publications Service

Office Of Science And Technology (NSF)
Research Organizations

Settings
use ENVIRONMENT

Settlements
use COMMUNITIES

Sex (Differences In Employment Or Compensation)
use EMPLOYMENT
MEN
WOMEN

Sex (Differences In Performance)
use PERFORMANCE
MEN
WOMEN

Shelf List
use LIBRARY CLASSIFICATION SYSTEMS
INDEXES

SIE (Science Information Exchange)
use SCIENCE AND TECHNOLOGY
INFORMATION SCIENCE SERVICE ORGANIZATIONS
RESEARCH

Sight
use VISION

SIGNALS
BT Symbol Sets
RT Body Language
Kinesics

Silence
use SOUND

Simulation
use MODELING

SINGING
BT Oral Communication
RT Dancing
Speech

Situation
use CONDITIONS

SIZE
  X   Length (Spatial Dimension)
  BT   Measures (Measuring Techniques)

SKILLS
  BT   Ability
  RT   Employment

SLA (Special Libraries Association)
  use LIBRARIES
       INFORMATION SCIENCE PROFESSIONAL
           ORGANIZATIONS

SLIDES
  BT   Audiovisual Software (Documents)

SMELL
  BT   Senses

Smoking (Behavior)
  use BEHAVIOR

Social Scientists
  use BEHAVIORAL SCIENCES
       SCIENTISTS

SOCIAL SECURITY
  BT   Compensation

Socially Handicapped
  use DISADVANTAGED PERSONS

Societies
  use ORGANIZATIONS

Society
  use CULTURES

SOCIOECONOMIC GROUPS
  X   Classes (Social)
       Middle Class
       Poor
  BT   Groups
  RT   Cultures
       Depressed Areas
       Ethnic Groups
  PX   Minority Groups

Software (Computer Programs)
use PROGRAMS

SORTING
BT Interrelations
RT Collating
Comparing
Electromechanical Data Processing Equipment
Merging
Selection
Sequencing

SOUND
X Acoustics
Noise (Sound)
Silence
BT Resources
RT Audiovisual Software (Documents)
Engineering
Equipment
Music (Art Form)
Nervous System
Oral Communication
Physical Sciences
Senses
Sound Recordings
Telephone
PX Audio Cassettes
Audio Input-Output
Indexing, Phonetic
Phonetic Indexing
Vowels

SOUND RECORDINGS
X Recordings (Sound)
BT Audiovisual Software (Documents)
RT Sound
PX Recorded Books

Source Indexing
use AUTHORS
INDEXING

Sources
use RESOURCES

Space (Allocated For Storage)
use STORAGE SPACE

Space (Outer Space)
  use AEROSPACE SCIENCES

SPACE (RESOURCE)
  BT Resources
  NT Storage Space
  RT Control Functions
     Measures (Measuring Techniques)
     Priorities

Space (Topologic)
  use TOPOLOGY

SPACE PROVISION
  BT Delivery Of Services

Space Utilization
  use STORAGE SPACE
     USES (OF RESOURCES)

SPAIN
  BT Europe

Spanish Americans
  use ETHNIC GROUPS

SPANISH LANGUAGE
  BT Natural Languages

Special Education
  use INSTRUCTING
     HANDICAPPED PERSONS

Special Libraries Association (SLA)
  use LIBRARIES
     INFORMATION SCIENCE PROFESSIONAL
        ORGANIZATIONS

SPECIFICATIONS
  X Design (Noun)
     Recommendations
     Requirements (Specifications)
  BT Criteria
  RT Bylaws
     Designing
  PX American National Standards Institute
        (ANSI)
     ANSI (American National Standards
        Institute)

Committee Z 39
Job Analysis
Job Description
Z39 Committee

SPEECH
Note: Used For Preparation, Organization And Delivery Of Speeches; Also For The Mechanics And Logistics Of Forming And Uttering Sounds Vocally.
X Phonation
Phonetics
BT Oral Communication
RT Linguistics
Multimedia Techniques
Nervous System
Singing
PX Automated Speech
Native Speakers
Vowels

SPEECHES (DOCUMENTS)
BT Documents
RT Meeting Documents
Oral Communication

SPEED
BT Measures (Measuring Techniques)

Stacks (Library Shelving)
use STORAGE SPACE

Staff (Employees)
use PERSONNEL

Standardization
use APPLICATIONS
CRITERIA

Standards
use CRITERIA

STATE GOVERNMENT
BT Governments

State-Of-The-Art Studies
use EVALUATION
REVIEWS

STATES
  BT United States Of America
  RT Geographic Areas

STATISTICS
  BT Mathematics
  NT Correlating
     Predicting
     Probability
     Sampling
     Variables
  RT Data Analysis
     Delphi Method
     Evaluation Techniques
     Game Theory
     Indicators
     Numeric Processing
     Research
     Techniques
  PX Bradford's Law
     Trends

Stimulation
  use APPLICATIONS
      STIMULI

STIMULI
  BT Input Process
  PX Stimulation

STORAGE AND RETRIEVAL PROCESSES
  X Document Retrieval
    Information Retrieval (Process)
    Mechanized Retrieval
    Retrieval (Process)
  BT Data Processing
  NT Inquiry Negotiation
     Searching
  RT Evaluation
     Information
     Selection
     Techniques
  PX Automated Retrieval
     Chemical Documentation
     Data Bases (For Automated Storage And Retrieval)
     Documentation (Processing Of Documents)

STORAGE AND RETRIEVAL SYSTEMS
X Data Banks
Information Retrieval (Systems)
Information Searching Systems
Information Systems
BT Information Science Service Organizations
NT Management Information Systems
RT Searching
PX Dial-Access Information Retrieval
Document Storage
Drug Information Systems
Hospital Information Systems
Industrial Information Systems
Information Collection (Noun) (Group Of Documents)
Information Collection (Noun) (Pool Of Data)
Law Enforcement Information Systems
MEDLARS Project (Medical Literature Analysis And Retrieval Systems)
Reservation Systems

STORAGE MEDIA
X Audiovisual Software (Media)
Cartridges
Cassettes
Computer Storage
Digital Storage
Drums (Memory)
Files (Media)
Paper
BT Equipment
NT Cards
Film (Photographic)
Magnetic Discs
RT Documents
Memory
Microform Documents
Multimedia Techniques
Preservation
Reproducing (Of Documents)
Reproducing Equipment
PX Continuous Records
Discontinuous Records
Mass Storage
Microfiche (Medium)
Random Access
Random Access Storage

Storage Medium Capacity
Unit Records
Vertical Files

Storage Medium Capacity
use CAPACITY
STORAGE MEDIA

STORAGE SPACE
X Space (Allocated For Storage)
Stacks (Library Shelving)
BT Space (Resource)
PX Capacity, Storage
Space Utilization

Strategy
use TECHNIQUES

STRENGTH
BT Measures (Measuring Techniques)

STRESS
BT Interaction

String-Processing Languages
use ARTIFICIAL LANGUAGES
WORD PROCESSING

STRUCTURE
X Directed Graphs
Form (Structure)
Syndetic Structure
BT Interrelations
NT Arrangement
Index Terms (Format And Interrelations)
RT Independence
Interaction
PX Boards (Of Governance)
Book Catalog
Card Catalogs
Information (Structure Of)
Language Structures
Lattices

Students
use LEARNING
PERSONNEL

Style (Writing)
  use WRITING (CREATIVE)

Style Manuals
  use WRITING (CREATIVE)
      BOOKS

Subject Analysis (of Documents)
  use INDEXING

Subject Cards
  use CARDS
      INDEXES

Subject Catalog
  use INDEXES

Subject Classification
  use CLASSIFICATION SYSTEMS

Subject Heading Lists
  use THESAURI

Subjects (of Documents)
  use INDEX TERMS (DOCUMENT SURROGATES)

Subjects (Research)
  use POPULATION

Subsidize
  use FINANCE

SUBSTANCES
  BT Materiel
  NT Drugs
      Food

SUBURBAN
      Note: Used For Either Geographic Location Or Activities Which Are Typically Suburban.
  BT United States Of America
  RT Geographic Areas
      Regional

Summaries
  use ABSTRACTS (OF DOCUMENTS)

Superimposed Coding
  use CODING SYSTEMS
      RANDOMNESS

Supervision
  use ADMINISTRATION

SURVEYING
  BT Data Gathering
  NT Interviewing
     Monitoring
  RT Diary Techniques
     Questionnaire Techniques
     Questionnaires
     Random Alarm Techniques
     Reviews
  PX User Studies (Surveying Users Of Services)

SURVEYS (RECORDS)
  BT Records

SWEDEN
  BT Europe

SWITCHING
  BT Transmission
  PX Referral

Switching Equipment
  use COMPONENTS

SWITZERLAND
  BT Europe

Syllabus
  use INSTRUCTING
      PROGRAMS

SYMBOL SETS
  X Characters
    Language Symbols
    Letters (Symbols)
  BT Languages
  NT Alphabets
     Body Language
     Coding Systems
     Kinesics
     Numbering Systems
     Signals

RT Linguistics
Printing
PX Character Recognition
Type Fonts
Words

SYMBOLIC LOGIC
BT Logic
RT Boolean Algebra
Linguistics
Mathematics

Symposia
use MEETINGS (COMMUNICATING VIA MEETINGS)

Syndetic Structure
use STRUCTURE

Synonyms
use SEMANTICS

Synopsis
use ABSTRACTS

SYNTAX
X Analysis, Syntactic
Articles (Parts Of Speech)
Grammar
Morpheme
Parsing
Phonemes
Sentence
BT Linguistics
PX Transformational Grammars

Synthesize
use CREATIVITY

System Analysis
use APPLICATIONS
EVALUATION

System Design
use DESIGNING
APPLICATIONS

System Operators
use PERSONNEL

Systems Phenomena
  use RELATIONS

Systems Theory
  use CYBERNETICS

T

Tables (Tabular Representation)
  use GRAPHIC RECORDS

Tabulating Cards
  use PUNCHED CARDS

TAPE
  BT Storage Media
  NT Magnetic Tape
     Paper Tape
     Videotape (Medium)
  PX Tape Reading Equipment

Tape Reading Equipment
  use READING EQUIPMENT
      TAPE

Tape Typewriters
  use INPUT-OUTPUT EQUIPMENT

TASTE
  BT Senses

Taxonomy
  use CLASSIFICATION SYSTEMS

Teachers
  use INSTRUCTING
      INFORMATION SCIENCE PERSONNEL

Teaching
  use INSTRUCTING

Teamwork
  use INTERACTION

Technical Writing
  use SCIENCE AND TECHNOLOGY
      WRITING (CREATIVE)

TECHNIQUES
  X   Approach
       Methods
       Procedures
       Strategy
  BT   Input
  NT   Communication Techniques
       Data Processing Techniques
       Evaluation Techniques
       Manual Techniques
       Photography
       Printing
  RT   Applications
       Delivery Of Services
       Input-Output Equipment
       Instructing
       Measures (Measuring Techniques)
       Modeling
       Publishing
       Research
       Statistics
         Storage And Retrieval Processes
  PX   ASCA (Automatic Subject Citation Alert)
           (Tradename)
       Automatic Subject Citation Alert (ASCA)
           (Tradename)
       Computer Assisted Literature Alerting
       DDD (Direct Distance Dialing)
       Direct Distance Dialing (DDD)
       Indexing Techniques
       Manipulative Indexing
       Multiprocessing
       Search Strategy

Technology
  use SCIENCE AND TECHNOLOGY

TELECOMMUNICATION
  X   Television Process
  BT   Communication Techniques
  NT   Radio
       Telemetry
       Television
  RT   Communication Satellites
       Microwave Techniques
       Multimedia Techniques

Telegrams
  use CORRESPONDENCE (DOCUMENTS)

TELEMETRY
  BT   Telecommunication

TELEPHONE
  BT   Input-Output Equipment
  PX   DDD (Direct Distance Dialing)
       Dataphone (Tradename)
       Dial-Access Information Retrieval
       Direct Distance Dialing (DDD)
       Picturephones
  RT   Oral Communication
       Sound

Teletypewriters
  use INPUT-OUTPUT EQUIPMENT

TELEVISION
  X   Cable Television (CATV)
       CATV (Cable Television)
       Closed-Circuit Television
       Color Television
  BT   Telecommunication
  RT   Cathode Ray Tubes
       News Publications
       Oral Communication
       Visual Communication
  PX   Educational Television (ETV)
       ETV (Educational Television)
       Kinescopes
       Mass Media
       Picturephones

TELEVISION EQUIPMENT
  BT   Audiovisual Equipment
  NT   Cathode Ray Tubes

Television Process
  use TELECOMMUNICATION

TEMPERATURE
  X   Cold (Temperature)
  BT   Physical Properties

Terminals (Computer)
  use COMPUTER TERMINALS

Terms (Indexing)
  use INDEX TERMS (DOCUMENT SURROGATES)

Testing
  use EVALUATION

Tests
  use MEASURES (MEASURING TECHNIQUES)

Text (Machine-Readable)
  use MACHINE READABLE DATA

Texts (Books)
  use BOOKS

THEORY
  X Hypotheses
    Null Hypothesis
  BT Concepts
  NT Game Theory
    Information Theory
    Learning Theory
    Queuing Theory
  RT Research
  PX Communication Theory
    Decision Theory
    Hypothesis Testing
    Indexing Theory

THESAURI
  X Authority Lists
    Controlled Vocabularies
    Lexicon (Controlled Language)
    Microthesaurus
    Query Languages (Conrolled)
    Retrieval Languages (Controlled)
    Search Languages (Controlled)
    Subject Heading Lists
    Vocabularies (Controlled Language)
  BT Documents
  RT Dictionaries
    Index Terms (Document Surrogates)
    Responses (To Inquiries)
    Searching
    Selection

Theses
  use DISSERTATIONS

Thinking
  use INTELLECT (INTELLIGENCE)

TIME
X Currency (Of Information)
Duration
Lag
Leisure
Length Of Time
Long Term
Permanent
BT Measures (Measuring Techniques)
NT Age (Of Persons Or Material)
Day
Future
Night
Year
RT Forecasts (Documents)
Predicting
Priorities
Scheduling
PX Current Awareness
Real-Time
Response Time

TIME-SHARING
X Multiplexing
BT Data Processing Techniques
RT Input-Output Equipment
Man-Machine Interaction
Online Data Processing
Priorities
Programs

Time-Sharing Networks
use TIME-SHARING SERVICES

TIME-SHARING SERVICES
X Bibliographic On-line Display (BOLD) (Tradename)
Communication Networks (Automated)
Computer Networks
Computer Utilities
Direct-Access Computing Techniques
Interactive Computer Services
Networks, Time-Sharing
Time-Sharing Networks
BT Information Science Service Organizations
RT Computer Applications
Online Data Processing
PX Arpanet

BALLOTS (Bibliographic Automation Of Large Libraries Using Time-Sharing)
BCN (Biomedical Communications Network)
Bibliographic Data Bases
Biomedical Communications Network (BCN)
EDUCOM (Educational Communications)
Information Networks (Automated)
Inter-University Communications Council

TOPOLOGY
X Space (Topologic)
BT Geometry

TOUCH
BT Senses

Tracings (Indexing)
use INDEX TERMS (FORMAT AND INTERRELATIONS)

Trainees
use LEARNING
PERSONNEL

Training
use INSTRUCTING

Transceivers (Transmitters/Receivers) (Equipment)
use TRANSMITTERS (EQUIPMENT)
RECEIVERS (EQUIPMENT)

Transcribing
use TRANSLATION

Transcription
use TRANSLATION

Transcripts
use RECORDS

Transformational Grammars
use TRANSLATION
SYNTAX
PROGRAMS

Transforming
use TRANSLATION

Transition
use CHANGE

Translater (Human)
  use LANGUAGE TRANSLATION
    INFORMATION SCIENCE PERSONNEL

TRANSLATION
  X Conversion
    Transcribing
    Transcription
    Transforming
    Transliteration
  BT Communication
  NT Coding
    Interpreting
    Language Translation
  RT Convertibility
    Word Processing
  PX Analog-Digital Conversion
    Analog-Digital Conversion Equipment
    Assembling (Computer Programs)
    Automated Translation (Transformation)
    Compiling (Computer Programs)
    Compression (Of Data)
    Converter
    Data Compression
    Data Conversion
    Data Mapping
    File Compression
    Interpreter (Equipment)
    Transformational Grammars

Transliteration
  use TRANSLATION

TRANSMISSION
  BT Communication
  NT Disseminating
    Switching
  RT Delivery Of Services
    Equipment
    Information Flow
    Routing
  PX Information Transfer
    Transmission Facilities

Transmission Facilities
  use TRANSMISSION
    EQUIPMENT

TRANSMITTERS (EQUIPMENT)

BT Input-Output Equipment
RT Disseminating
PX Data Transmission
Facsimile Transmission
Transceivers (Transmitters/Receivers) (Equipment)

Transmitters (Human)
use COMMUNICATION
INFORMATION SCIENCE PERSONNEL

TRANSPORTATION
X Delivery (Transport Of Goods)
BT Processes (Processing)
RT Motor Vehicles

TRAVEL
BT Behavior
NT Visits
RT Disseminating
PX Trip Reports

Tree Structures
use CLASSIFICATION SYSTEMS

Trends
use EVALUATION
STATISTICS

Trip Reports
use TRAVEL
RECORDS

Trucks
use MOTOR VEHICLES

TURKEY
BT Europe

Tutorial Documents
use INSTRUCTING
DOCUMENTS

Type Fonts
use PRINTING EQUIPMENT
SYMBOL SETS

Typewriters
use PRINTING EQUIPMENT

Typography
  use COMPOSING

U

Understanding
  use KNOWLEDGE (STATE OF BEING INFORMED)

Unification
  use MERGING

Union Catalogs
  use INDEXES
    COOPERATION

Union Lists
  use INDEXES
    COOPERATION

Unisist
  use INTERNATIONAL (GEOGRAPHY)
    INFORMATION SCIENCE SERVICE ORGANIZATIONS

Unit Records
  use STORAGE MEDIA
    ARRANGEMENT

UNITED KINGDOM
  X Great Britain
  BT Europe
  NT Ireland
  PX ASLIB (Association of Library And Information Bureaux)
    Association of Library And Information Bureaux (ASLIB)
    BNB (British National Bibliography)
    British (People)
    British National Bibliography (BNB)

UNITED STATES AIR FORCE
  BT Department Of Defense
  RT Military Science

UNITED STATES ARMY
  BT Department Of Defense
  RT Military Science

UNITED STATES CONGRESS
  X   Congress (U.S.)
  BT   United States Government

United States Food And Drug Administration (FDA)
  use PUBLIC HEALTH SERVICE

UNITED STATES GOVERNMENT
  X   Federal Government
       National Government (U.S.)
  BT   Governments
  NT   Central Intelligence Agency (CIA)
       Department Of Commerce
       Department Of Defense
       Environmental Protection Agency (EPA)
       Health, Education And Welfare Department (HEW)
       National Aeronautic And Space Administration (NASA)
       National Bureau Of Standards
       National Referral Center
       National Science Foundation
       Office Of Education
       United States Congress
  RT   Libraries
       National (Nationwide)
  PX   CFSTI (Clearinghouse For Federal Scientific And Technical Information)
       Civil Service
       Clearinghouse For Federal Scientific And Technical Information (CFSTI)
       FBI (Federal Bureau Of Investigation)
       Federal Bureau Of Investigation (FBI)
       Federal Information Processing Standards
       General Services Administration (GSA)
       GSA (General Services Administration)
       National Technical Information Service (NTIS)
       NTIS (National Technical Information Service)
       Peace Corps

UNITED STATES NAVY
  BT   Department Of Defense
  RT   Military Science

UNITED STATES OF AMERICA
  BT   Geographic Areas

NT Cities
Communities
Depressed Areas
National (Nationwide)
Regional
Rural
States
Suburban

UNITED STATES PATENT OFFICE
BT Department Of Commerce

UNIVERSAL DECIMAL CLASSIFICATION SYSTEM
BT Library Classification Systems

Universities
use ACADEMIC ORGANIZATIONS

UNIVERSITY AND COLLEGE EDUCATION
X Graduate Education
Higher Education
Professional Training
BT Education

University Libraries
use ACADEMIC ORGANIZATIONS
LIBRARIES

Update
use ERROR DETECTION AND CORRECTION

Upgrade
use CHANGE

Urban
use CITIES

URBAN GOVERNMENT
BT Governments
RT Cities

Use (Of Information Resources)
use USES (OF RESOURCES)

User Needs
use NEED
USERS (OF SERVICES)

User Studies (Surveying Users Of Services)

use SURVEYING
USERS (OF RESOURCES)

USERS (OF RESOURCES)
Note: Used For Persons Who Are Themselves Providers Of Information, Such As Authors, Publishers, Or Librarians, As Contrasted With Persons Who Are Clientele Of Information Services.
X Information Users (Authors, Service Providers)
BT Information Science Personnel
RT Consulting Services
PX User Studies (Surveying Users Of Services)

USERS (OF SERVICES)
Note: Used For Persons Who Are The Clientele Of Particular Libraries, Information Services, Or Data Processing Services.
X Information Users (Information System Clientele)
Readers (Library Users)
BT Consumers
PX Index Instructions (To Users Of An Index)
User Needs
User Studies (Surveying Users Of Services)
Viewers (Persons Engaged In Visual Communication)

USES (OF RESOURCES)
Note: Used For The Process Of Using Resources or For Characterizing Such Use.
X Exploitation
Information Uses
Use (Of Information Resources)
BT Administration
NT Applications
Disposal
Maintenance
RT Compensation
PX Bradford's Law
Information Crisis
Research Utilization
Space Utilization

Utilities

use SERVICE ORGANIZATIONS

Utility
use EFFECTIVENESS

V

Validation
use CONTROL FUNCTIONS

VALUES
X Morals
BT Goals
NT Benefit
Criteria
Ethics
Priorities
RT Expectations
Motivation

Vandalism
use BEHAVIOR
DYSFUNCTIONS

VARIABLES
X Dependent Variables
Independent Variable
BT Statistics
RT Research
Sampling

Verification
use ERROR DETECTION AND CORRECTION

Vernacular
use NATURAL LANGUAGES
GEOGRAPHIC AREAS

Vertical Files
use STORAGE MEDIA
ARRANGEMENT

VIDEOTAPE (MEDIUM)
BT Tape

VIDEOTAPES (DOCUMENTS)
BT Audiovisual Software (Documents)

PX Electronic Videorecording (EVR)
EVR (Electronic Videorecording)

Viewers (Equipment)
use READING EQUIPMENT

Viewers (Persons Engaged In Visual Communication)
use USERS (OF SERVICES)
VISUAL COMMUNICATION

VISION
X Sight
BT Senses
RT Visual Communication
PX Optical Aids

VISITS
BT Travel

VISUAL COMMUNICATION
X Optical Display
BT Communication
NT Reading
RT Audiovisual Software (Documents)
Body Language
Dancing
Fine Arts
Kinesics
Multimedia Techniques
Oral Communication
Television
Vision
PX Viewers (Persons Engaged In Visual Communication)

Vocabularies (Controlled Language)
use THESAURI

Vocabularies (Natural Language)
use DICTIONARIES

VOLUME
BT Measures (Measuring Techniques)

VOLUNTARY SERVICES
BT Service Organizations
RT Volunteers

VOLUNTEERS
  BT Roles (Social)
  RT Paraprofessionals
     Voluntary Services

Vowels
  use SPEECH
      SOUND

W

WAGES
  X Income
    Overtime
    Pensions
    Salaries
  BT Compensation

Weeding (Of Document Collections)
  use DOCUMENTS
      DISPOSAL

WEIGHTING
  BT Probability
  RT Bias
  PX Index Terms, Weighting

WOMEN
  X Female
  BT Human Beings
  PX Sex (Differences In Employment Or
       Compensation)
     Sex (Differences In Performance)

WORD PROCESSING
  BT Data Processing
  RT Computer Applications
     Data Processing Services
     Data Processing Techniques
     Input-Output Equipment
     Language Translation
     Linguistics
     Numeric Processing
     Translation
  PX String-Processing Languages

Word Processing Equipment
  use   DATA PROCESSING EQUIPMENT

WORD PROCESSING SERVICES
  BT   Data Processing Services

Words
  use   SYMBOL SETS
        LANGUAGES

Work
  use   EMPLOYMENT

WORK (ENERGY OUTPUT)
  BT   Measures (Measuring Techniques)
  RT   Effectiveness
       Efficiency
       Employment

Work Performance
  use   PERFORMANCE

Work Study Program
  use   EMPLOYMENT
        LEARNING
        PERSONNEL

Workmen's Compensation
  use   COMPENSATION

Workshop (Meeting)
  use   MEETINGS (COMMUNICATING VIA MEETINGS)

WRITING
  X   Handwriting
  BT   Recording
  NT   Compiling
  RT   Writing (Creative)

WRITING (CREATIVE)
  X   Creative Writing
       Style (Writing)
  BT   Creativity
  RT   Publishing
  PX   Scientific Writing
       Style Manuals
       Technical Writing

X

Xerography (Tradename)
  use ELECTROSTATIC PROCESSES

Y

YEAR
  BT Time

YOUNG ADULTS
  BT Adults
  RT Children

Z

Zator Cards
  use EDGE-NOTCHED CARDS

Z39 Committee
  use INFORMATION SCIENCE SERVICE ORGANIZATIONS
      SPECIFICATIONS

HIERARCHIC ARRANGEMENT OF THE THESAURUS

INFORMATION SCIENCE

I. ENVIRONMENT
- CONDITIONS
  - CRISIS
  - FLOODS
- GOALS
  - EXPECTATIONS
  - MOTIVATION
    - INTERESTS
  - VALUES
    - BENEFIT
    - CRITERIA
      - INDICATORS
        - INDEX TERMS (DOCUMENT SURROGATES)
      - SPECIFICATIONS
    - ETHICS
    - PRIORITIES
- RELATIONS
  - BARRIERS
  - HUMAN RELATIONS
    - INTERPERSONAL RELATIONS
    - SATISFACTION
  - INDEPENDENCE
    - PRIVACY
  - INTERNATIONAL RELATIONS
- RESOURCES
  - FACILITIES
    - BUILDINGS
      - HOSPITALS
      - PRISONS
      - RESIDENTIAL FACILITIES
  - GEOGRAPHIC AREAS
    - AFRICA
    - ASIA

```
Environment
 (Resources)
   (Geographic Areas) continued
            CHINA
            INDIA
            ISRAEL
            JAPAN
            KOREA
            PAKISTAN
          AUSTRALIA
          CANADA
          EUROPE
            CZECHOSLOVAKIA
            DENMARK
            FRANCE
            GERMANY
            HUNGARY
            ITALY
            NORWAY
            POLAND
            RUSSIA
            SPAIN
            SWEDEN
            SWITZERLAND
            TURKEY
            UNITED KINGDOM
              IRELAND
          INTERNATIONAL (GEOGRAPHY)
          LATIN AMERICA
          MEXICO
          UNITED STATES OF AMERICA
            CITIES
            COMMUNITIES
            DEPRESSED AREAS
            NATIONAL (NATIONWIDE)
            REGIONAL
            RURAL
            STATES
            SUBURBAN
        HUMAN RESOURCES
          ABILITY
            CREATIVITY
              DESIGNING
              MODELING
              WRITING (CREATIVE)
            SKILLS
          HUMOR
          INTELLECT (INTELLIGENCE)
```

Environment
- (Resources)
  - (Human Resources)
    - (Intellect) continued
      - COGNITION
        - CONCEPTS
          - THEORY
            - GAME THEORY
            - INFORMATION THEORY
            - LEARNING THEORY
            - QUEUING THEORY
        - LEARNING
        - REASONING
        - RECOGNITION
          - PATTERN RECOGNITION
      - KNOWLEDGE (STATE OF BEING INFORMED)
      - PERSONALITY
        - ATTITUDES
    - KNOWLEDGE AREAS
      - ARTS AND HUMANITIES
        - FINE ARTS
          - ARCHITECTURE
          - LITERATURE
          - MUSIC (ART FORM)
          - SCULPTURE
      - MATHEMATICS
        - ALGEBRAS
          - BOOLEAN ALGEBRA
        - GEOMETRY
          - TOPOLOGY
        - LOGIC
          - SYMBOLIC LOGIC
          - STATISTICS
            - CORRELATING
            - PREDICTING
            - PROBABILITY
              - RANDOMNESS
              - WEIGHTING
            - SAMPLING
            - VARIABLES
      - PHILOSOPHY
      - SCIENCE AND TECHNOLOGY
        - BEHAVIORAL SCIENCES
          - ANTHROPOLOGY
          - ECONOMICS
            - BANKING
          - EDUCATION
            - ELEMENTARY SCHOOL EDUCATION

- Environment
  - (Resources)
    - (Knowledge Areas)
      - (Science and Technology)
        - (Behavioral Sciences)
          - (Education) continued
            - HIGH SCHOOL EDUCATION
            - POSTGRADUATE EDUCATION
            - UNIVERSITY AND COLLEGE EDUCATION
          - HISTORY
          - LAW
          - LIBRARY SCIENCE
            - BIBLIOGRAPHY (PROCESS)
            - ABSTRACTING
            - INDEXING
              - COORDINATE INDEXING
              - KWIC INDEXING
          - LINGUISTICS
            - SEMANTICS
            - SYNTAX
          - MILITARY SCIENCE
          - POLITICAL SCIENCE
          - PSYCHOLOGY
            - BEHAVIOR
              - CONDITIONING
              - PERFORMANCE
              - RECREATION
              - RESISTANCE (HUMAN)
              - RESPONSIBILITY
              - TRAVEL
                - VISITS
        - BIOLOGICAL SCIENCES
          - AGRICULTURE
          - BIOMEDICAL SCIENCES
            - HEALTH CARE
              - DENTISTRY
            - NEUROLOGY
              - NERVOUS SYSTEM
                - SENSES
                  - HEARING
                  - SMELL
                  - TASTE
                  - TOUCH
                  - VISION
          - ECOLOGY
          - GENETICS
        - CHEMISTRY
          - CHEMICAL COMPOUNDS

Environment
- (Resources)
  - (Knowledge Areas)
    - (Science and Technology) continued
          - PHYSICAL SCIENCES
            - AEROSPACE SCIENCES
            - ATOMIC AND MOLECULAR PHYSICS
            - CYBERNETICS
            - ELECTRONICS
            - ENGINEERING
            - GEOLOGY
            - METALLURGY
        - LANGUAGES
          - ARTIFICIAL LANGUAGES
            - ALGOL (PROGRAMMING LANGUAGE)
            - APL (PROGRAMMING LANGUAGE)
            - BASIC (PROGRAMMING LANGUAGE)
            - COBOL (PROGRAMMING LANGUAGE)
            - COURSE WRITER (PROGRAMMING LANGUAGE)
            - FORTRAN (PROGRAMMING LANGUAGE)
            - INTERLINGUA
            - LISP (PROGRAMMING LANGUAGE)
            - MUMPS (PROGRAMMING LANGUAGE)
            - PL/1 (PROGRAMMING LANGUAGE)
          - NATURAL LANGUAGES
            - ARABIC LANGUAGE
            - CHINESE LANGUAGE
            - ENGLISH LANGUAGE
            - FRENCH LANGUAGE
            - GERMAN LANGUAGE
            - JAPANESE LANGUAGE
            - RUSSIAN LANGUAGE
            - SPANISH LANGUAGE
          - SYMBOL SETS
            - ALPHABETS
              - BRAILLE
            - BODY LANGUAGE
              - DANCING
            - CODING SYSTEMS
            - KINESICS
            - NUMBERING SYSTEMS
            - SIGNALS
      - LIGHT
      - MATERIEL
        - PERSONAL PROPERTY
        - PHYSICAL PROPERTIES
          - TEMPERATURE
        - SUBSTANCES

Environment
 (Resources)
  (Materiel)
   (Substances) continued
        DRUGS
        FOOD
      MEMORY
      POPULATION
       ANIMALS
       HUMAN BEINGS
        GROUPS
         CULTURES
         DISADVANTAGED PERSONS
         ETHNIC GROUPS
         HANDICAPPED PERSONS
         SOCIOECONOMIC GROUPS
        MEN
        WOMEN
      SOUND
      SPACE (RESOURCE)
       STORAGE SPACE
     <u>ROLES (SOCIAL)</u>
      ADULTS
       AGED
       YOUNG ADULTS
      APPLICANTS
      AUTHORS
      CHILDREN
      CLIENTELE
       USERS (OF SERVICES)
      CONSUMERS
      FAMILY
      MIGRANTS
      PATIENTS
      PEERS
      PERSONNEL
       INFORMATION SCIENCE PERSONNEL
        USERS (OF RESOURCES)
       PARAPROFESSIONALS
       SCIENTISTS
      REPRESENTATIVES
      VOLUNTEERS
     <u>RULES</u>
      LAWS
       BYLAWS
       COPYRIGHT LAW
       COURTS
      POLICIES
      RIGHTS

II. FEEDBACK AND CONTROL
- ADMINISTRATION
  - CONTROL FUNCTIONS
    - CENSORSHIP
    - CHANGE
      - GROWTH
    - DECISION-MAKING
      - PLANNING
        - DESIGNING
    - IDENTIFICATION
  - EMPLOYMENT
    - COMPENSATION
      - AWARDS
      - INSURANCE (LIFE)
      - RETIREMENT PLANS
      - SOCIAL SECURITY
      - WAGES
  - FINANCE
    - ACCOUNTING
    - BUDGETING
    - COSTS
    - FINANCIAL AID
      - GRANTS AND CONTRACTS
    - MARKETING
      - ADVERTISING
      - EXHIBITING
  - INTERACTION
    - ACCESSING
    - ADJUSTMENT
    - AVAILABILITY
    - BROWSABILITY
    - COMPATIBILITY
    - CONVERTIBILITY
    - COOPERATION
      - EXCHANGING
        - INTERLIBRARY LOAN
    - COORDINATION
    - IMPROVEMENT
    - INFLUENCING
    - MAN-MACHINE INTERACTION
    - REDUNDANCY
    - RISK
    - SECURITY
    - STRESS
  - INTERRELATIONS
    - ASSOCIATING
      - BROWSING
    - BIAS
    - COLLATING

Feedback and Control
(Interrelations) continued

- COMPARING
- COMPOSING
  - PHOTOCOMPOSITION
- CONSISTENCY
- CONSTRUCTING
- EXTRACTING
- FILE ORGANIZATION
- MEETING ORGANIZATION
- MERGING
- OPPORTUNITY
- PERMUTATION
- RANKING
- READINESS
- SEQUENCING
  - SCHEDULING
- SORTING
- STRUCTURE
  - ARRANGEMENT
    - CLASSIFICATION SYSTEMS
      - FACETED CLASSIFICATION SYSTEMS
      - LIBRARY CLASSIFICATION SYSTEMS
        - DEWEY DECIMAL CLASSIFICATION SYSTEM
        - LIBRARY OF CONGRESS CLASSIFICATION SYSTEM
        - UNIVERSAL DECIMAL CLASSIFICATION SYSTEM
    - INDEX TERMS (FORMAT AND INTERRELATIONS)

USES (OF RESOURCES)
- APPLICATIONS
  - COMPUTER APPLICATIONS
  - OPERATION (OF EQUIPMENT)
- DISPOSAL
- MAINTENANCE
  - PRESERVATION
  - PREVENTION
  - RELEASE
  - RESTORATION
  - RETIREMENT

<u>EVALUATION</u>
- DIAGNOSIS
- MEASURES (MEASURING TECHNIQUES)
  - CAPACITY
  - EFFECTIVENESS
  - EFFICIENCY

- Feedback and Control
  - (Evaluation)
    - (Measures) continued
      - FEASIBILITY
      - FREQUENCY
      - INCIDENCE
      - PRECISION
      - PRODUCTIVITY
      - QUALITY
      - QUANTITY
      - RECALL/PRECISION (MEASURES)
      - RELIABILITY
      - SATISFACTORINESS
      - SIZE
      - SPEED
      - STRENGTH
      - TIME
        - AGE (OF PERSONS OR MATERIAL)
        - DAY
        - FUTURE
        - NIGHT
        - YEAR
      - VOLUME
      - WORK (ENERGY OUTPUT)
    - RESEARCH
      - ARCHIVAL RESEARCH
      - EXPERIMENTAL RESEARCH
        - DEMONSTRATION
      - METHODS RESEARCH
      - OBSERVATIONAL RESEARCH
      - OPERATIONS RESEARCH

- III.INPUT
  - <u>EQUIPMENT</u>
    - AUTOMATA
    - COMMUNICATION SATELLITES
    - COMPONENTS
      - CIRCUITS
      - LENSES
    - DATA PROCESSING EQUIPMENT
      - COMPUTERS
        - ANALOG EQUIPMENT
        - CALCULATORS
        - ELECTROMECHANICAL DATA PROCESSING EQUIPMENT

Input
(Equipment)
(Data Processing Equipment)
(Computers) continued
ELECTRONIC DIGITAL COMPUTERS
MICROPROCESSORS
MINI-COMPUTERS
HYBRID COMPUTERS
INPUT-OUTPUT EQUIPMENT
AUDIOVISUAL EQUIPMENT
TELEVISION EQUIPMENT
CATHODE RAY TUBES
COMPUTER TERMINALS
LIGHT PEN
PRINTING EQUIPMENT
GRAPHICS PRINTING EQUIPMENT
READING EQUIPMENT
OPTICAL SCANNERS
RECEIVERS (EQUIPMENT)
REPRODUCING EQUIPMENT
TELEPHONE
TRANSMITTERS (EQUIPMENT)
INSTRUMENTS
FORMS (SURVEY)
QUESTIONNAIRES
LASERS
MOTOR VEHICLES
STORAGE MEDIA
CARDS
PUNCHED CARDS
EDGE-NOTCHED CARDS
FILM (PHOTOGRAPHIC)
MAGNETIC DISCS
TAPE
MAGNETIC TAPE
PAPER TAPE
VIDEOTAPE (MEDIUM)
INFORMATION
DATA
DEMOGRAPHIC DATA
MACHINE READABLE DATA
DOCUMENTS
ABSTRACTS (OF DOCUMENTS)
ABSTRACTS PUBLICATIONS
ADVERTISING LITERATURE
AUDIOVISUAL SOFTWARE (DOCUMENTS)
GRAPHIC RECORDS
BLUEPRINTS

Input
 (Information)
  (Documents)
   (Audiovisual Software)
    (Graphic Records) continued
          FLOW DIAGRAMS
          PAINTINGS
          PHOTOGRAPHS
            HOLOGRAMS
        MICROFORM DOCUMENTS
          MICROFICHE DOCUMENTS
          MICROFILM DOCUMENTS
          MICROPRINT DOCUMENTS
        MOTION PICTURES
        MUSIC (DOCUMENTS)
        SLIDES
        SOUND RECORDINGS
        VIDEOTAPES (DOCUMENTS)
      BIBLIOGRAPHIES
        CITATIONS
      BIOGRAPHICAL DOCUMENTS
      BOOKS
      CORRESPONDENCE (DOCUMENTS)
      DICTIONARIES
      DISSERTATIONS
      DOCUMENTARIES
      EXHIBITS (DOCUMENTS)
      FORECASTS (DOCUMENTS)
      IDENTIFICATION (DOCUMENTS OF)
      INDEXES
        CITATION INDEXES
        KWIC INDEXES
      LAY PUBLICATIONS
      MEETING DOCUMENTS
      NAME AND ADDRESS LISTS
      NEWS PUBLICATIONS
      PAMPHLETS
      PATENTS
      PERIODICALS
      PROPOSALS
      PUBLICATIONS (PERSONAL OR
        ORGANIZATIONAL)
      RECORDS
        INPUT RECORDS
        MANUSCRIPTS
        MEDICAL RECORDS
        OUTPUT RECORDS
          SAMPLES

```
Input
 (Information)
  (Documents) continued
            RESEARCH RESULTS
            SURVEYS (RECORDS)
          REPRODUCTIONS
          REVIEWS
          SPEECHES (DOCUMENTS)
          THESAURI
      INPUT PROCESS
        ACQUIRING
        INQUIRIES
        STIMULI
      LIBRARIES
        LIBRARY OF CONGRESS
        NATIONAL AGRICULTURAL LIBRARY
        NATIONAL LIBRARY OF MEDICINE
      MUSEUMS
      NEED
        INFORMATION NEED
      ORGANIZATIONS
        ACADEMIC ORGANIZATIONS
        GOVERNMENTS
          STATE GOVERNMENT
          UNITED STATES GOVERNMENT
            CENTRAL INTELLIGENCE AGENCY (CIA)
            DEPARTMENT OF COMMERCE
              CENSUS BUREAU
            UNITED STATES PATENT OFFICE
            DEPARTMENT OF DEFENSE
              DEFENSE DOCUMENTATION CENTER (DDC)
              DEFENSE INTELLIGENCE AGENCY
              UNITED STATES AIR FORCE
              UNITED STATES ARMY
              UNITED STATES NAVY
            ENVIRONMENTAL PROTECTION AGENCY (EPA)
            HEALTH, EDUCATION AND WELFARE
              DEPARTMENT (HEW)
              PUBLIC HEALTH SERVICE
            NATIONAL AERONAUTIC AND SPACE
              ADMINISTRATION (NASA)
            NATIONAL BUREAU OF STANDARDS
            NATIONAL REFERRAL CENTER
            NATIONAL SCIENCE FOUNDATION
            OFFICE OF EDUCATION
            UNITED STATES CONGRESS
          URBAN GOVERNMENT
        MANUFACTURING AND SALES ORGANIZATIONS
```

- Input
  - (Organizations) continued
      - PHILANTHROPIC ORGANIZATIONS
      - PROFESSIONAL ORGANIZATIONS
        - INFORMATION SCIENCE PROFESSIONAL ORGANIZATIONS
      - SERVICE ORGANIZATIONS
        - INFORMATION SCIENCE SERVICE ORGANIZATIONS
          - STORAGE AND RETRIEVAL SYSTEMS
            - MANAGEMENT INFORMATION SYSTEMS
          - TIME-SHARING SERVICES
          - VOLUNTARY SERVICES
    - <u>TECHNIQUES</u>
      - COMMUNICATION TECHNIQUES
      - MICROWAVE TECHNIQUES
      - MULTIMEDIA TECHNIQUES
      - PROGRAMS
      - TELECOMMUNICATION
        - RADIO
        - TELEMETRY
        - TELEVISION
      - DATA PROCESSING TECHNIQUES
        - OPTICAL COINCIDENCE TECHNIQUES
        - PROGRAMMING
        - TIME-SHARING
      - EVALUATION TECHNIQUES
        - DATA ANALYSIS
          - FACTOR ANALYSIS
        - PERT (PROGRAM EVALUATION REVIEW TECHNIQUES)
      - DATA GATHERING
        - CRITICAL INCIDENT METHOD
        - DELPHI METHOD
        - DIARY TECHNIQUES
        - QUESTIONNAIRE TECHNIQUES
        - RANDOM ALARM TECHNIQUES
        - SELECTION
        - SURVEYING
          - INTERVIEWING
          - MONITORING
    - MANUAL TECHNIQUES
    - PHOTOGRAPHY
      - HOLOGRAPHY
    - PRINTING

IV. OUTPUT
- <u>DYSFUNCTIONS</u>
  - DETERIORATION
  - FAILURE
  - NOISE (IN COMMUNICATION SYSTEMS)
- <u>NORMAL FUNCTIONS</u>
  - RESPONSES
    - DELIVERY OF SERVICES
      - AUDIOVISUAL PRODUCTION SERVICES
      - BINDING SERVICES
      - COMMUNICATION SERVICES
      - CONSULTING SERVICES
        - COUNSELING
      - COPYING SERVICES
      - DATA PROCESSING SERVICES
        - WORD PROCESSING SERVICES
      - DELIVERY OF DOCUMENTS
      - EDITING SERVICES
      - EDUCATIONAL SERVICES
      - EQUIPMENT PROVISION
      - LANGUAGE TRANSLATION SERVICES
      - PRINTING SERVICES
      - PUBLISHING SERVICES
      - RESPONSES (TO INQUIRIES)
      - SPACE PROVISION

V. PROCESSES (PROCESSING)
- <u>COMMUNICATION</u>
  - INSTRUCTING
    - COMPUTER ASSISTED INSTRUCTING (CAI)
      - PLATO SYSTEM
    - COURSES (EDUCATIONAL)
    - CURRICULA
    - SELF-INSTRUCTING
  - ORAL COMMUNICATION
    - MEETINGS (COMMUNICATING VIA MEETINGS)
    - SINGING
    - SPEECH
  - PERSON-TO-PERSON COMMUNICATION
  - PUBLISHING
  - RECORDING
    - WRITING
      - COMPILING
  - TRANSLATION

- Processes
  - (Communication)
    - (Translation) continued
        - CODING
        - INTERPRETING
        - LANGUAGE TRANSLATION
      - TRANSMISSION
        - DISSEMINATING
          - INFORMATION FLOW
          - ROUTING
          - SDI
        - SWITCHING
      - VISUAL COMMUNICATION
        - READING
    - DATA PROCESSING
      - BATCH PROCESSING
      - ERROR DETECTION AND CORRECTION
      - NUMERIC PROCESSING
      - OFFLINE DATA PROCESSING
      - ONLINE DATA PROCESSING
      - STORAGE AND RETRIEVAL PROCESSES
        - INQUIRY NEGOTIATION
        - SEARCHING
      - WORD PROCESSING
    - REPRODUCING (OF DOCUMENTS)
      - ELECTROSTATIC PROCESSES
      - PHOTOCOPYING
    - TRANSPORTATION

SINGLE WORD INDEX TO MULTI-WORD TERMS

ABSTRACTS	ABSTRACTS (OF DOCUMENTS)
	ABSTRACTS PUBLICATIONS
ACADEMIC	ACADEMIC ORGANIZATIONS
ADDRESS	NAME AND ADDRESS LISTS
ADMINISTRATION	NATIONAL AERONAUTIC AND SPACE ADMINISTRATION (NASA)
ADULTS	YOUNG ADULTS
ADVERTISING	ADVERTISING LITERATURE
AERONAUTIC	NATIONAL AERONAUTIC AND SPACE ADMINISTRATION (NASA)
AEROSPACE	AEROSPACE SCIENCES
AGE	AGE (OF PERSONS OR MATERIAL)
AGENCY	CENTRAL INTELLIGENCE AGENCY (CIA)
	DEFENSE INTELLIGENCE AGENCY
	ENVIRONMENTAL PROTECTION AGENCY (EPA)
AGRICULTURAL	NATIONAL AGRICULTURAL LIBRARY
AID	FINANCIAL AID
AIR	UNITED STATES AIR FORCE
ALARM	RANDOM ALARM TECHNIQUES
ALGEBRA	BOOLEAN ALGEBRA
ALGOL	ALGOL (PROGRAMMING LANGUAGE)
AMERICA	LATIN AMERICA
	UNITED STATES OF AMERICA
ANALOG	ANALOG EQUIPMENT
ANALYSIS	DATA ANALYSIS
	FACTOR ANALYSIS
APL	APL (PROGRAMMING LANGUAGE)
APPLICATIONS	COMPUTER APPLICATIONS
ARABIC	ARABIC LANGUAGE
ARCHIVAL	ARCHIVAL RESEARCH
AREAS	DEPRESSED AREAS
	GEOGRAPHIC AREAS

CHAPTER 1

FUNCTIONS OF A THESAURUS

As can be seen from Diagram A (p.2), a thesaurus is used to standardize terminology when indexing documents* for potential retrieval; it is also used to index inquiries so that matching between documents (represented by their indexing terms) and inquiries (represented by their indexing terms) can be achieved. A thesaurus is used to control language used by a retrieval system. The way it does this is to greatly reduce (use only a subset of) the number of words available in a natural language. It also restricts the grammar that may be used, in that thesaurus terms are primarily nouns, gerunds or verbs; they are rarely adjectives, participles, or adverbs, and never connectives such as articles or prepositions, unless the latter are contained in a phrase that is a thesaurus term (a term can consist of more than one word). A thesaurus sometimes defines terms arbitrarily to conform to the subject matter of the retrieval system, for example "cell (biology)" or "cell (battery)" thus excluding other possible meanings of "cell" such as "cell (prison)." Another way thesauri may control use of language is through rules about how terms may be combined. This topic is presented later, in the discussion of subordinate (heading-subheading) indexing.

---

*documents is used throughout this book to mean all written records, such as books, journals, microforms, reports, sound recordings, videotapes, or magnetic tapes.

A "system" view of a thesaurus is depicted in Diagram A. It is patterned after suggestions in a paper by Barbara Kyle, in 1964[1].

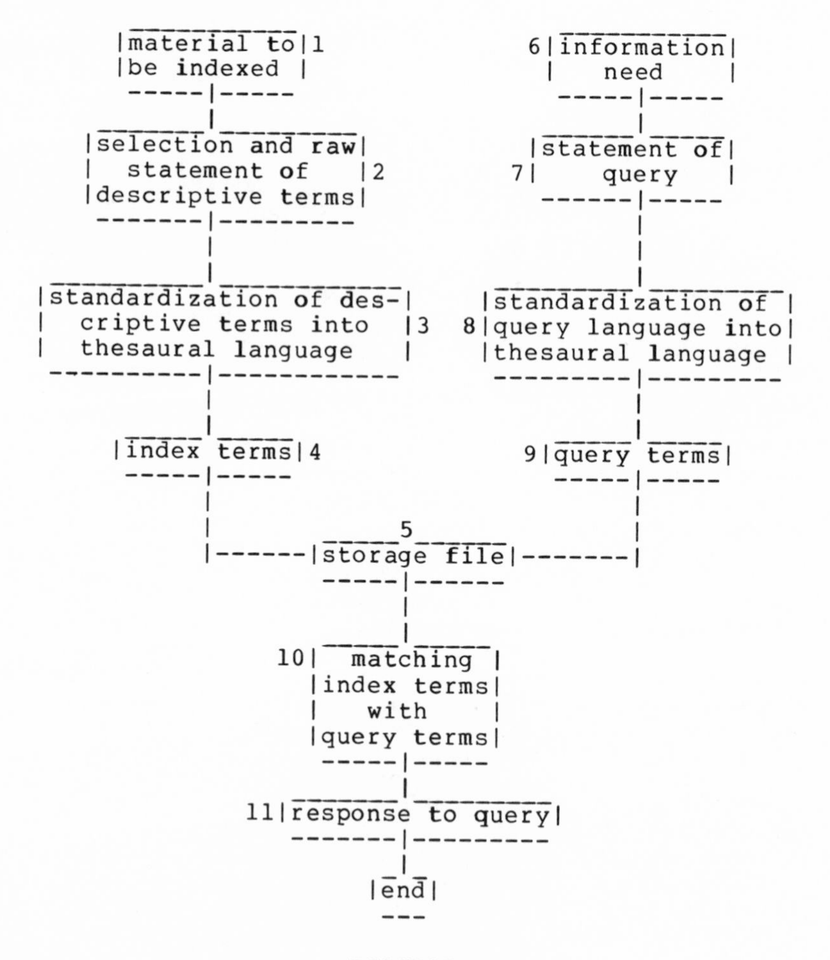

DIAGRAM A

FUNCTIONS OF A THESAURUS

The alternative to controlled retrieval-system language, via a thesaurus, is use of free (natural) language. An example of uncontrolled indexing is a concordance, or its variant, a KWIC (Keyword-In-Context) index [2]. In these, almost every text word other than connectives is made available as an index word, without so much as standardizing singular and plural forms of the same word. The in-context technique reduces ambiguity (over that of single-word concordances) by providing phrases which, for example, distinguish homographs:

...testing of zinc and lead.
...his turn to lead the troops,

distinguish areas of interest:

...water quality of the Mississippi.
...political views of Mississippi democrats,

or differences of phrasing between two authors:

...sittin' on de fence.
..sitting on the fence.

Sometimes this last example is used persuasively to demonstrate that uncontrolled language is highly desirable for retrieval systems because it protects precise phrasing of authors, may disclose cultural, psychological or legalistic differences, etc. However, there are persuasive counter arguments having to do with costs and speed of search. Free language systems, because of their richness of expression (many ways to say the same thing) prove quite diffuse when it comes to searching them for concepts rather than particular words or phrases. The searcher faces the task of having to guess all of the ways a concept may have been expressed by authors. Experience has shown that searchers cannot expect to win the guessing game every time, and in most cases search strategies become highly complex and/or lengthy.

This is mentioned because in recent decades strong consideration has been given to use of free language indexing and searching, with the thought that computers make possible searching that human beings find too tedious or time-consuming[3,4,5,]. Persons who have examined KWIC indexes know that they must be searched from A to Z to be certain that all interesting (pertinent) phrasing of the concept being searched has been located. However, if the concordance were kept within the computer's memory, the computer, given a thesaurus or a search formula which specified all of the related terms, could do that searching and print out for human inspection only the references which matched the query. As was pointed out, it requires imagination on the part of the searcher and a complex search statement to anticipate the major word combinations which might be used to express a concept; for example:

bed
sofa
couch
day-bed
divan
cot
recliner
etc....

to express "furniture on which to lie down." Suppose the search concept were: "birds of flight" or "vehicles of transportation"? Not only is the searching difficult and expensive in most cases; the storage of large files of text and the updating of those files are also expensive at today's state of technology. Full-text searching can be very useful in cases where thesaurus standardization or indexer error have caused needed search items to be unavailable, as in the following cases:

1. The thesaurus says: "Cot use BED," but the query specifies that only documents containing the word "cot" are of interest. In a thesaurus system, this requires that all documents retrieved under the word "bed" be screened to find the ones that actually mention "cot." In such cases, perhaps it would be less expensive to be able to do a full-text search using the word "cot"? One can only speculate

because there are many factors to consider.

2. Suppose an indexer for some reason failed to index one or more of the pertinent documents even with the term "bed"? Here it would clearly be useful to have a back-up system of some sort, especially a full-text one, if it were important to obtain total document recall from the system, as one might for legal purposes.

Considerations such as the above, while important, lead away from the more central efficiency and effectiveness considerations, such as how most queries can be answered at an acceptable level of comprehensiveness and acceptable cost. For most retrieval systems, where the subject-oriented queri tend to be of the nature:

1. What are the most recent documents about X?
2. Did anyone ever do X?
3. Who are the leading authors in field X? etc....

it is usually less expensive to prepare for the required searching by 1) indexing documents via a controlled vocabulary (thesaurus), 2) storing indexing terms along with the identification of the document to which they pertain and 3) searching that store by indexing the query, using the same standardization tool (thesaurus) as was used for indexing the documents that are to be retrieved. In the best of all retrieval worlds, the search armamentarium would consist of not only a thesaural system but a full-text system, a citation system, and perhaps various others to be used interchangeably and synergistically. But if given a choice of only one of these, for either manual or automated use, in the majority of cases it seems wise to choose a thesaural system.

REFERENCES

1. Kyle, Barbara R. F., Information Retrieval and Subject Indexing: Cranfield and After. J. Doc. 20:55-69 (June) 1964.

2. Luhn, H. P., Keyword-in-context Index for Technical Literature (KWIC Index) Ann. Doc. 11 (4): 288-95 (Oct.) 1960.

3. Horty, John F., Exercise with the Application of Electronic Data Processing Systems In General Law: Proceedings of the Electronic Data Retrieval Committee of the Bar Activities Section of the Amer. Bar Assoc. Washington D.C. Aug. 29, 1960, pp. 3-4

4. Andrews, Donald D. Experience with Electronic Searching of United States Patents. Proceedings of the Electronic Data Retrieval Committee of the Amer. Bar. Assoc. Washington D. C. Aug. 29, 1960, pp. 15-20.

5. International Business Machines (IBM) STAIRS program for information storage and retrieval.

CHAPTER 2

HOW TO CONSTRUCT A THESAURUS

Who and What to Use as Authorities

Thesaurus-building, like most activities, can be approached either theoretically or empirically. The empiricists among thesaurus builders examine language use by actual (randomly selected) authors within the subject field of concern, and make many of their content decisions statistically; decisions such as whether to include particular words or phrases, and if so, whether they should be "accepted" terms or "use" terms, are made according to how often they occur in sampled documents, and how much overlap there is in their usage. For example, if given data from a sample of documents such as the following:

	occurrences
demonstrating employability	14
developing employability	3
improving employability	10
measuring employability	4
predicting employability	6
becoming employable	2
becoming employed	80
becoming employees	15
achieving employment	2
returning to employment	70
training for employment	60
.	
.	
.	
vocational rehabilitation	450

one needs to decide whether all five words:

employability
employable
employed
employees
employment

will be incorporated into the thesaurus, and whether some will be made "use" terms. When doing this, one needs to decide whether becoming employees and becoming employed are synonyms; how to distinguish between becoming employable and becoming employed, if at all; whether vocational rehabilitation includes employment training, or vice versa, and so on. One also has to decide whether there are intellectual and statistical grounds for creating two-word terms such as:

employability measurement
employment training.

The empirical approach makes explicit and compelling the decisions which have to be made, and it presents enough evidence of usage to guide those decisions. It may also protect against errors such as forgetting that a term sometimes has different meanings, for example:

manual (book)	manual (method)
painting (object)	painting (process)
hearing (legal)	hearing (physiological)
cheese cake (food)	cheese cake (pose)

or that certain abbreviations are better known than full spellings, for example:

DDT
polio
UNESCO.

By contrast, the theoretical approach to thesaurus-building begins much as Melvil Dewey did (these days we use thesaural committees) by attempting to think "logically" about how the universe should be divided and expressed. Usually, such thinking is based on an hierarchical examination of knowledge rather than examination of language use--and the results are certain to be different in character. For one thing, biases (experiences) of the person(s) concerned will be reflected in the outcome. It is said that Dewey did not believe in the use of alcohol or tobacco, so he neglected (originally) to make a place in his scheme for books on these subjects. Members of thesaural committees can be expected to behave in terms of their own views and interests, and this can cause a great deal of unproductive argument, or else arbitrary conformance to an authoritarian view. Usually, the fewer persons involved in producing a thesaurus, the faster the work will go, and this fact argues for use of empirical data to counteract biases, oversights and other errors or limitations of thesaurus-builders.

The foci of activity within the field for which the thesaurus is being built may be quite different from what the theoretician may imagine; therefore, the theoretically conceived thesaurus is almost certain to be less efficient for indexing and searching than one based on the reality of statistical plus logical thinking. The theoretically derived thesaurus may be more stable in the long run, but stability is not the most desirable property of a thesaurus. To work well, a thesaurus must be a living, dynamic tool, and must change as activities and word usage change. This fact greatly increases the difficulty and complexity of thesaurus maintenance.

Somewhere between the theoretical and empirical approaches, lies the use of published authorities from which to glean terms and on which to depend for definitions or selectiveness. It is prudent in any case to seek out and examine all previous attempts at thesauri for the field of concern, so that one does not risk reinventing the wheel. It is also prudent to learn as much as possible from published indexes, glossaries and other existing guides to the content of the field. But to rely on scattered works to the extent of copying or clipping terms and trying to form an amalgam from a number of sources is something of an exercise in futility. What is achieved by so doing tends to be a hodge-podge.

Empirical Construction Procedures

Because this thesaurus builder believes so strongly in the empirical approach to thesaurus-building, procedures presented here will assume the willingness and capability to obtain necessary start-up data; that is, evidence of what creators of the literature concerned are talking about, with what priority and frequency, and what words or word strings they use to express their messages. It also is assumed that preliminaries such as checking for existence or other thesauri, and analyzing them, have been completed.

Another important start-up pursuit is to examine the qualifications of thesaurus-builder(s) and, depending on the size of the project, try to find one or more experienced persons with a good grasp of the subject matter with which the thesaurus will deal. Technique can be as important as subject background, if not more so, whenever extensive contextual analyses of usage can be depended on for guidance. Contextual analysis requires examining words as they are imbedded in sentences or fragments of sentences, so that nuances of meaning, homographs, suffixes, and other linguistic aspects of words and groups of words can be examined. As was mentioned earlier, context helps with decisions about whether a term should consist of more than one word, as in "heart rate," "citric acid cycle," "mobile homes," and the like. And, it helps with decisions about accepting, perhaps, both noun and verb (gerund) forms of a word, such as both "ships" and "shipping" or "employment" and "employees," should statistical as well as logical clues indicate the concepts to be differentiated by and important to users of the language under consideration.

The size of the sample to be analyzed statistically, linguistically, and logically varies, of course, with the number of documents to be indexed and searched, and with the rate of change of language use within the field. A new field or one that is expanding rapidly, such as virology a few decades ago, or particle physics at present, will need to be examined more carefully than one which is relatively stable. A one-tenth sample of the first year's document supply (for the storage and retrieval system that is to ensue) is almost always ample, if one wants to be one's own statistician when determining sample size. Should a large collection be the indexing target, the expense of the analysis dictates a more scientific approach to sampling.

Many computer service centers can supply the analog of a KWIC (Keyword-In-Context)[2] index to titles or summaries of documents, through use of extant computer programs which can be operated rather inexpensively, especially if the text to be analyzed is already in machine-readable form (keyed onto punched cards, paper or magnetic tape) or can be made so by using an optical scanning device. If the text to be analyzed must be keyed especially for the purpose, cost is considerably higher.

Readers who might become involved in producing KWIC listings should be advised to retain at least two words on either side of the word being examined, to the extent possible within a single sentence, as is shown in some of the examples that follow.

Connectives such as "and," "for," "of," and "to" can be prevented (through the computer program) from occurring in the alphabetic window of the KWIC listing (Table A). This is so one will not produce page after page of printout of the word "and" with contextual words on either side of it. If connectives are prevented from occurring in the window position, the program can still retain the connectives as context for other words, as is shown also in Table B. Perhaps a more compelling example is the following:

the young	woman	and child
injured	women	with child

where "and" and "with" play an important semantic role, and obviously need to be visible in order to be considered.

Once the in-context list is in hand, it is read from A to Z to gain an overview of language use. Each different word encountered in the alphabetic window is considered, to recognize first how many different contexts there are for it, then how many variants it has and whether it seems to need subdivision, as:

	WINDOW	
and health	rehabilitation	of workers
vocational	rehabilitation	for hundreds
concerning the	rehabilitation	process
provision of	rehabilitation	facilities by

Table A

IN-CONTEXT WORD LIST

Both mental and written notes are made during this first perusal of the in-context list, which will begin to relate like terminology and to discriminate synonyms and terms of choice. The second time through the list large slips (5"x8" or larger) should be made to begin recording decisions about each word encountered. Consider a KWIC listing with the following appearance:

1	from his	abilities	to perform	
2	to various	abilities	deriving from	
3	without any	abilities	which allow	
4	judging from	abilities	and capacities	accept plural form
1	trust the	ability	of the aged	
2	respect the	ability	to accomplish	
3	to special	ability	testing for	
4	reading	ability	and speech	
5	commonly used	ability	measurements	
6	effect on	ability	to achieve	

TABLE B

Based on overall statistics for the sample, let us

assume that either the term "ability" or "abilities" should be included in the thesaurus. Thus, the note "accept plural form," was made during the first inspection of the list. If the thesaurus is to be used by a computer program for automatic spelling checks or perhaps standardization of "raw" indexing terms, there may need to be a thesaurus entry for both the singular and plural forms, since computers are not able to recognize (without resorting to word stems) that "ability" is closely related to "abilities." If so, one slip to be made will read as follows:

Ability use ABILITIES.

And the slip for the plural form will read:

ABILITIES
refer from: Ability.

During the manual thesaurus construction, the reciprocal note ("Refer from," in this case) is always made immediately following any decision, wherever in the alphabet an affected workslip may be located, and even if one has to construct a workslip just for the purpose of recording a reciprocal. It is not necessary at this time to consider broader and narrower terms or other terms relating to "ABILITIES"; these considerations will follow, as the systematic thesaurus-building technique runs its course. At this point it is important only to record the decision just made, so that if it is changed any and every affected work slip can also be changed; otherwise one quickly creates chaos and must start again, since starting over is usually easier than unravelling a residue of unsystematic threads of thought, perhaps tangled several times over.

CHAPTER 3

THESAURI VERSUS AUTHORITY LISTS

Since neither a thesaurus nor an authority list has a well known definition, arguments about whether or how they differ tend to occur and to be very muddy. The purpose of this section is to try to define each of them sufficiently so that they can be compared, in the hope that this will smooth the way to better communication and understanding among information-retrieval colleagues, some of whom have not obtained training in library science.

The best known of all thesauri is no doubt that of Roget[6]. Roget's purpose was to help writers or speakers to choose among many synonyms or related phrasings, so they might use the most precise term for expressing an intended meaning. As was explored in Chapter 1, an information-science thesaurus would seem to do the opposite; it reduces richness of expression so that many ways of expressing a concept are transformed to a single standardized expression (however imprecise it may seem). With a little imagination, the difference between Roget's thesaurus and other information science thesauri can be recognized as one of emphasis; they are the inverse of one another. This will become clear later in this chapter, when discussion turns to the note structure of an information science thesaurus. Or, it can be seen in the body of this "Thesaurus of Information Science Terminology" that the "Refer from" note gathers terms that are synonyms (and synonym-like) to make explicit the expressions which a particular thesaurus term represents. Thus, Roget's intended function can be served as well as its inverse.

"Authority list" is a term well known to librarians, but to few others. If one views information science as including library science, there were authority lists in the field of information science long before there were thesauri. It is believed that the linguist Masterman[7] was the first after Roget to use the term "thesaurus" to express the idea of mapping relationships among words and phrases which address the same concept. "Subject heading list" is used more or less synonymously with "authority list" by librarians, because subject heading lists were components of library classification systems such as those devised by Melvil Dewey[8] and the Library of Congress (LC)[9]. These classification schemes, through their alphabetic indexes, became the authority for headings and heading-subheading combinations used in the card catalogs of various libraries.

Subject heading lists are essential components of hierarchical classification schemes. They give the user an alphabetic key to the hierarchy--otherwise it would be more difficult for students to learn to catalog books during cataloging classes or for a person in a cataloging department of a library to follow in a predecessor's footsteps, because all novices would be left on their own to locate (somewhat ambiguously and interpretively) the "right" place in the hierarchy, and how to phrase the meaning of that "right" place. For example in LC:

> "Insulin shock" is included in "RC661.16 Diabetes-Special Therapies." If one were indexing a publication on insulin shock, it could not be found within the Library of Congress classification schedule directly. Through systematic scanning of the hierarchic scheme, one might encounter the note: "RCL660 Diabetes Mellitus and other disorders of the pancreatic internal secretion," which has been updated through "RC661.16 Diabetes-Special Therapies." Even so, this requires knowing a priori that insulin has a relation-

ship to diabetes and that it is a therapeutic agent. Accessing the LC classification scheme through its Subject Heading List refers the reader from insulin to diabetes and states the class number under which to locate "diabetes." Not only is it more direct to be told the appropriate class number than to have to discover it; one also learns that "insulin shock" translates into "diabetes-special therapies" so far as assigning an index heading is concerned.

Among specialists in the matter, it is generally recognized, now, that the intended purposes of thesauri and authority lists are identical--in accord with Diagram A in Chapter 1--even though formats and uses may not be. We will turn first to differences in format, examining the parts of a traditional classification system such as those of Dewey or LC. Each divides knowledge into intellectual segments, then subdivides them repeatedly, as finely as seems useful for the retrieval system to be served.

One can speculate that the appropriate degree of fineness at libraries such as the LC is judged by the number of books to be "marked and parked" (an expression coined by Robert Fairthorne[10]) within any particular subdivision. Because atomic energy was unknown at the time either the Dewey or LC classification was devised, of course no space was created for it. In the 1950's, a suitable place for atomic energy had to be created, and it was often the case that the classification number became too long to fit on the spine of a book.

The point is that multiple indexing and retrieval functions are served by the three components of any classification system:

1) an hierarchic arrangement of concepts, expressed primarily in language which indicates the logic of their placement; but this language is not necessarily to be used verbatim for subject headings.

2) An alphabetically arranged compilation of words and phrases needed for indexing

(creating subject headings for) books, keyed to the pertinent hierarchic placement(s) of the concepts concerned. This is known as the subject heading list for the classification scheme and serves as an index to it, but the subject heading list also serves as the authority (standard) (controlled vocabulary) for indexing and retrieving documents. As in other retrieval systems, the terms of a query made of the system must be made to conform with subject headings, in order to achieve a match.

3) A system of numbers which correspond to the divisions and subdivisions of the hierarchy, and which are used for marking and parking the books. This function is not needed in many retrieval systems because the documents may be organized (instead of by classification number) according to author, or accession number, or some other parking system.

As in the case of Roget versus other information science thesauri, emphasis placed on the components of a classification system may be an obscuring factor impeding direct comparison of authority lists (subject heading lists) with other thesauri. The term "classification system" indicates that the main concern when the system was created was with the marking and parking of the documents. To obtain the classification numbers required, the hierarchy was a prerequisite. Thus these two components of the system, partially because of their size as well (the classification schedule is often much more voluminous than the subject heading list), seemed to influence the librarian's thinking in the direction that the subject heading list was secondary, perhaps only an appendage for convenient access to the hierarchy. Even the term "cataloging," which should mean a unique discursive description of library-held documents, so that each can be distinguished from the others via the card catalog, has been used sometimes to mean only assignment of classification numbers, so that each book will have a unique shelf address.

During the cataloging process, as taught in most library schools, selection of subject headings for a book is done only after classification, number assignment and descriptive cataloging are completed, usually in that order. Thus, novice librarians are taught to concentrate on technical processing of books before considering retrieval of them by library clientele. As a library school student, I remember cataloging the Chemical Rubber Company's "Handbook of Chemistry and Physics," and my instruction from the teacher can be summarized something like this:

> "Notice the three nouns in the title. Because 'handbook' belongs within stated subject areas, one inverts handbook after the subject, as 'Chemistry--handbooks.' Because 'Chemistry' occurs before 'Physics' in the title, 'Chemistry' becomes the subject heading of choice. Yes, one also could make a subject card for 'Physics--handbooks,' but it is not necessary."

Thus, if authority is achieved through many different rules and procedures, including use of subject heading lists, and the card catalog is thought of as a mechanism only for books, it is difficult for librarians to project these procedures onto retrieval systems for such materials as the Publication Board Reports[11] (confiscated scientific writings, originally on microfilm) of the post-World War II period, or for journal articles or medical records. In the United States, persons who had to deal with these other kinds of documents found no help from librarians during the 1940's and early 1950's, and had to invent the indexing procedures for themselves. They first identified themselves as "documentalists" based on the precedent set by those in the European "Federation Internationale de Documentation." This group of documentalists later called themselves "information scientists." They were mostly men, had not been to library school, and were not on speaking terms with librarians, but they had the functions depicted in Diagram A, Chapter 1 to fulfill. They did not need conventional library classification numbering systems, because they typically organized documents by serial numbers, but they did have a very definite need for standardized terminology for indexing and searching. There was great fuss and ferment over how to achieve retrieval by means of manual or mechanical card systems and later, magnetic tape systems; whether to use single words or phrases, or a combination of the two, for indexing and searching; and whether alphabetic or hierarchic arrangements of terminology were better, etc. It eventually came to the point where librarians and information scientists sat down together under the encouragement of the Z39 Committee of the American National Standards Institute and produced a "Guide to Thesaurus Construction and Use"[12]. This Guide may have been the first published consensus that the purpose of the librarian's authority list and of the information scientist's thesaurus was identical. The Subcommittee found that even the note structures were surprisingly similar, and that indeed there did have to be both an hierarchic and alphabetic arrangement of the concepts and terminology, no matter what the end product was called. The term "thesaurus" was chosen over "authority list" because the Z39 subcommittee's consensus was that information scientists concern themselves with all types of recorded materials, including correspondence, engineering drawings, journal articles, recordings and many other types of materials not necessarily found in libraries. Information science thesauri are constructed to enable storage and retrieval systems for any or all of these materials (more readily than if authority lists, coupled to library classification schemes are used).

This discussion can now devote itself briefly to the matter of content of thesauri, concentrating on the note structure of thesauri versus authority lists, according to Table C, which was originally constructed for the first edition of this thesaurus in 1966[13]. The authority lists' "see" became "use" in early thesauri, and "see also" became "related terms." Their reciprocals "see from" and "refer from" are almost uniformily "refer from" in thesauri. "See also specific" translates into "narrower terms" in thesauri.

The non-comparable notes are shown also in

Table C. "May be subdivided as" has no counterpart in thesauri. "Narrower term" and "broader term" have no counterparts in authority lists. The introduction to this "Thesaurus of Information Science Terminology" explains the "px," "partial refer from," note, used in this thesaurus.

	Authority List	Thesaurus
Comparable Notes:		
	See:	Use:
	See Also:	Related Terms
	See Also Related:	Related Terms
	x (See From:)	Refer From
	Refer From	Refer From
	xx (See Also From:)	Related Term
	Scope Note:	Scope Note
	Use:	Scope Note
	See Also Specific:	Narrower Term
Non-Comparable Notes:		
	May Be Subdivided As:	*************
	****************	Broader Term
	****************	px
	****************	Partial Refer From

TABLE C

REFERENCES

6. Roget, Peter M. Roget's Thesaurus of English Word and Phrases. Ed. R. A. Deutch. New York, New York, St. Martin's Press, 1965.

7. Masterman, Margaret M. Cambridge Language Research Unit, England. Statement based on a personal communication between this author and H. P. Luhn. See reference to the Luhn-Masterman relationship in Schultz, Claire K. ed., H. P. Luhn: Pioneer of Information Science. New York, New York, Spartan, 1968, p. 266.

8. Dewey Decimal Classification and Relative

Index. 3 vols. Albany, New York, Forest Press, 1971.

9. Library of Congress Classification Schedules and Subject Heading List. Currently published by Gale Research Co., Detroit.

10. Fairthorne, Robert A. Towards Information Retrieval. London, Butterworth, 1961.

11. Publication Board Reports, U.S. Government Publications of "confiscated" Scientific Documents from Axis Countries.

12. American National Standards Association. Guidelines for Thesaurus Construction and Use. 1972, p. 34.

13. Schultz, Claire K., compiler and editor, Information Science Thesaurus, Phila., Pa. Drexel Institute of Technology, Nov. 1964, p. 120.

CHAPTER 4

USE OF A THESAURUS FOR INDEXING

The alphabetic arrangement of the thesaurus is usually the arrangement of greatest use to the indexer, partially because the "use" notes are of considerable importance. The reader will remember that a complete thesaurus has two primary arrangements--alphabetic and hierarchic. It may have others, such as an alphabetic index to the words contained in multiword terms, as this thesaurus does.

Given a particular document to index, an indexer begins the task with:

1) the document, containing the word choices of the particular author(s),

2) the indexer's personal background in the subject matter of the document; and perhaps some prior knowledge of the content of the storage and retrieval system for which the indexing is being done and

3) the storage and retrieval system's thesaurus, which records (through constant updating) every "accepted" indexing term which the system recognizes, and every rule with regard to using that term (scope notes, etc.). In addition, the thesaurus contains every rule for dealing with non-accepted terms (through "use" references) which have been made explicit for that system.

As a tutorial device, in what follows we will pursue the hypothetical experience of a novice indexer being introduced to a storage and retrieval system. Suppose she is in a new job and is indexing the first document ever assigned to her. Assume she has some knowledge of the subject matter concerned, or she would not have been hired for the indexer position.

Because of pressures of time, indexers in general tend to rely on a document's title and abstract for discovering the document's message. Let us assume our indexer has begun by reading these, and by quickly scanning the document as a whole. Titles typically have three or four "keywords" (usually nouns or noun phrases) upon which the document's abstract and its sub-titling arrangement will have elaborated sufficiently for the indexer to have good understanding of what each keyword in the title means. Therefore, it is reasonable to begin the indexing process by trying to match those keywords with the content of the thesaurus, using the thesaurus's alphabetic arrangement.

Hopefully, the first keyword being looked up will appear in its expected position; if so, it will either be an "accepted" term or a "use" term. If it is an accepted term, the indexer needs to ascertain that the scope note, or lack of one, makes the term acceptable in the context in which it occurs in the document. For example, if the word being looked up is "cell" and there are thesaurus entries for:

CELLS (BIOLOGICAL)
CELLS (ELECTRICAL)
CELLS (PLASMA)
CELLS (PRISON)

the indexer must choose appropriately, transcribing the parenthesized modifier (this is one kind of scope note) as part of the chosen term. For clarity, let's digress a moment to say that there is another kind of scope note which human indexers, in contrast to current computer indexing programs, can use; for example:

ELECTRICITY
Note: reserved for general works on the subject. Choose more specific terms whenever possible.

In this case, the indexer should consult the "narrower terms" note for "electricity" to decide whether narrower terms will do. It is a general rule in indexing to use the most specific terms possible, but this does not necessarily preclude using, in addition, somewhat broader terms, if they seem useful for retrieval purposes, and are in keeping with the rules of the thesaurus. Individual storage and retrieval systems vary in their practices regarding the use of both a specific term and its parent term (hierarchically) for indexing the same document, so one must abide by those rules as they are encountered.

Getting on with thesaurus use for indexing the document at hand, let us assume that the first title keyword with which the indexer chose to work, indeed, found a thesaurus match and that it was an accepted term, within scope. (If there is no scope note, then there is no decision to be made about scope; the term can be used for indexing regardless of context). For the second keyword, let us assume a thesaurus match, but one that took a little more ingenuity. For example, suppose the title contained a noun phrase such as "cw transmission" and there was no explanation of what "cw" abbreviated. Also, "cw" is not in the expected place in the thesaurus, and not noted as a type of transmission when "transmission" is consulted in the thesaurus.

Since the abbreviation is unknown to the indexer, this calls for consultation of an appropriate general dictionary (in this case, a dictionary of electrical or radio terms) so that it can be learned that "cw" stands for "continuous wave", the type of transmission used in sending Morse Code. Sure enough, the thesaurus contains the term "continuous waves" and the indexer wonders why there was not a cross-reference from "cw" to "continuous waves" and why there was no indication in the notes for "transmission" that "continuous waves" is either specific to or else related to "transmission." Knowing the rules of thesaurus construction, the indexer immediately pencils into her personal copy of the thesaurus the remedies for this oversight, and writes a note to her supervisor to bring it to more general attention.

To index the phrase "cw transmission" the indexer must use the two terms, "continuous waves" and "transmission," assuming a coordinate indexing system. As was mentioned in discussing authority lists for subject headings, thesauri can be used for controlling articulated (pre-coordinated) indexing systems, too. Had the hypothetical system being discussed not been a coordinate one, the indexer might have had to locate "transmission--continuous waves" as a single thesaurus term, or else have decided to use "transmission" as the most specific term available. However, even in an articulated system, "continuous waves" and "transmission" might happen to occur without subdivision, so the indexing result in such a case would be identical with that of the coordinate system. The slipperiness of this concept is illustrated by "continuous waves" being a two-word term, and therefore, in a sense, pre-coordinated; a pure coordinate system (one that coordinates at search time rather than indexing time) would use only one-word terms: in this case "continuous" and "waves."

From the foregoing, the set of indexing terms for the document tells us the document has something to do with "Cells (electrical)" and "transmission" by "continuous waves." Perhaps those are enough index terms? The answer depends on whether or not the content of the document is completely summarized by those terms. We suspect not, because one wonders what the relationship of electrical cells and continuous wave transmission might be.

Suppose it turned out that the full title of the document was "Cell Life of Batteries Used for CW Transmission from Marine Vessels." Hm... Maybe the indexer should have tried for "cell life" rather than "cells" as the thesaurus equivalent for the first key phrase? No, looking up "cell life" does not give any new indexing information. Does the thesaurus contain the term "batteries"? Even if it does, would it be useful to be able to search for "batteries" and "continuous waves" and "transmission" in order to retrieve the document being indexed? Assuming that our joint answer to the last question is yes, then "batteries" needs to be looked up in the thesaurus. Suppose it is an accepted term and its "broader term" note says that "power supply" is also a thesaurus term? Should "power supply" or "battery" or both be chosen as index terms? Looking up "power supply," the indexer encounters a scope note saying to use more specific terms whenever possible. The thesaurus resolves the matter: it eliminates the ambiguous situation while confirming the general rule that the most specific term should be used. But we are still convinced that both "batteries" and "cells (electrical)" are needed as indexing terms, so "batteries" is added to the indexing set for the document.

The indexer has the feeling that "marine vessels" or "ships" or "boats" or some such term should also be indexed, since marine conditions are different from land conditions, and the indexer has a nagging feeling that the concept of battery life or cell life needs to be more completely expressed than through the terms already in the indexing set. Looking up "ships" produces a thesaurus match, but while reading its notes, the indexer notices that "marine conditions" happens to be a related term. Being a novice, and wanting to play it safe, the indexer consults "marine conditions." She finds that for the document at hand it is appropriate to use this term because the scope note reads: "Note: used when a combination of air, water, temperature, etc. conditions is meant; otherwise use more specific terms." "Marine conditions" is added to the indexing set for the document, but what about "life"? "Life" cannot be found in the thesaurus; however "life span (biology)" is there, and among its "related terms" is "duration." That might do. Consulting "duration" the indexer finds among its related terms, "continuous use." After considering it, the indexer rejects "continuous use" because there is no indication in the document that cw transmission took place continuously. More probably use of the radio on the test vessels (the text of the document uses "boats," not the title word "vessels") was intermittent and therefore the batteries were at rest a good portion of the time. "Duration" is added to the index set, and now all of the concepts in the <u>title</u> seem to be represented by indexing terms.

Is the indexing set complete? The answer depends on how representative of the document the title is. In this case the title was informative, but there are probably a few additional concepts to be expressed. Is the document a bonafide report of research results? If so, the term "research" should be added (assuming it is in the thesaurus). Were the "marine conditions" encountered in various temperature zones or only one? In any case, perhaps "temperature" should be added? And so on....

Thus, we see how a relatively well-constructed thesaurus and a thinking-indexer might interact to good effect. Being a novice, the indexer expects feedback from a revisor of her work, to explain any errors of omission or commission during the first week or so on the job; after that, the thesaurus must either be or <u>become</u> sufficient to accomodate thoroughly and appropriately the concepts expressed by both authors of documents entering the storage system and by authors of inquiries put to the retrieval system (one and the same system).

Uses of the Various Arrangements of the Thesaurus

In addition to the uses just illustrated, a thesaurus helps to keep indexing consistent among various indexers for the same storage and retrieval system, and also helps a single indexer stay relatively consistent from day to day. It has been shown that indexing can be consistently bad, good, or indifferent; however, for dependable retrieval, indexing has to be good--and a necessary ingredient of good indexing is consistency. This will become clearer when we discuss search strategy in Chapter 5.

Earlier we noted that indexers use the alphabetic arrangement of the thesaurus for the most part. On occasion they also use the hierarchic arrangement or the single-word index to multiword terms. The following will illustrate:

1) when the novice indexer was wondering about whether to use "power supply", "batteries" and "cells (electrical)" as companion indexing terms, it might have been useful to consult the hierarchy under "electricity" and thereby, in a single lookup, determine whether the terms in question represented three different hierarchic levels, what their sibling terms were at each level, etc. The logic on which the thesaurus is based is often clarified by means of the hierarchy. Paranthetically, indexing a single document at three different hierarchic levels is usually counterindicated.

2) Suppose the indexer wondered if there were terms which contained the word "cell," but which did not have cell as their first word, and therefore could not be located alphabetically. Terms like "primary cell," or "secondary cell" might be encountered because they were listed in the notes for "cells (electrical)" but there were other possibilities such as "primary cell longevity" which would not be directly under "cells (electrical)" hierarchically, and might not even be listed as a related term for cells. The single-word index to multiword terms provides an easy way to deal with such thesaurus-related questions.

Note: The hypothetical document and thesaurus chosen for the above discussion might have illustrated information science terminology and made use of this thesaurus. In a live teaching situation, that would have been the wiser choice, because interaction among teacher, student, thesaurus, and document could have been ad hoc, and carried to exhaustion. In the non-live situation, given a real thesaurus, a document of his or her own choosing, and no one with whom to interact, the reader would have been at a disadvantage, because of terminology observed in the thesaurus but not explained in the tutorial discussion. However, with the foregoing presentation freshly in mind, the reader might now wish to choose an information science document, and attempt to index it by means of this thesaurus, guided by this thesaurus' set of "Instructions to Users." Such an exercise will reinforce the points brought out in this chapter.

CHAPTER 5

USE OF A THESAURUS FOR SEARCHING

Background

The previous chapter illustrated interaction between an indexer and a thesaurus. A searcher interacts with a thesaurus in much the same way. We noted that in indexing a document, the indexer's background knowledge and analytic ability were important ingredients of the indexing process. We might also have noted that the author's ability to organize and express concepts clearly in the title, abstract, and document as a whole affected the indexing process, and probably its outcome.

Searching involves 1) expression of the inquiry, 2) background and skill of any intermediary (the person who actually performs the search), 3) formulation of a search strategy with which to approach a storage and retrieval system and 4) the process of matching the strategy to various parts of the system.

All of the above aspects of searching are multifaceted, but let us confine ourselves here to pursuing a few of the facets of the matching process, mentioned last. At the current state-of-the-art, it is typical to match query terms with indexing terms (in manual indexes or in computer data bases) and then refer to the full documents for which references are retrieved, for either of two purposes:

1) decreasing the initial search's recall by culling for the most pertinent documents, or

2) increasing the recall by either:

   a) becoming aware of additional terminology useful for further searching, or else

   b) locating additional references through the bibliographies of the retrieved documents.

Of course, both the increasing and decreasing activities can be follow-on procedures, and if an intermediary is doing the searching, consultation with the inquirer can also be part of an iterative search process.

We are aware that in building the search strategy, the prior outcome of the indexer's interactions with the thesaurus must be dealt with. We noted in Chapter 4 that a thesaurus helps indexers to be consistent, but 100 percent consistency cannot be expected, so among other things, the search strategy has to compensate for possible indexing inconsistencies.

Additionally, just as the author of a document has to say what is meant, for good results the author of an inquiry has to be precise about what is wanted or needed. In reality, some inquirers state precisely what they want; some state precisely what they need (there is an important difference between want and need) but most inquirers make (often subconsciously) a somewhat global or diffuse inquiry, with the idea of refining it once they have received a response. It is for this reason that it is best for the inquirer to be actively involved in the formulation of the search strategy, and in the iterative stages of the matching process, whenever a search is delegated to an intermediary. Usually both persons benefit from the liaison and the search results are both less voluminous and more pertinent to the question.

Formulating A Hypothetical Strategy

Given imprecision in the inquiry, considerable latitude for incompleteness and ineptitude in the indexing, and also imperfections in the thesaurus, the wise searcher approaches search strategy formulation warily. The thesaurus is usually the prime tool for getting started, and again, the searcher usually goes first to the alphabetic arrangement, in order to benefit from the "use" references, if there are any pertinent ones. If the alphabetic arrangement proves unhelpful, or if the subject of the inquiry is an unfamiliar one, the searcher might want to peruse the hierarchy to learn the established relationships among them. Not all "use" references are necessarily logical or obvious; certainly the terms referenced are not always synonyms of the term being referred from. For example, in this thesaurus, we find:

concept coordination	use	COORDINATE INDEXING
concordances	use	INDEXES
conferences	use	MEETINGS
continuous records	use	ORGANIZATION; STORAGE MEDIA

One purpose of "use" references is to provide instructions, even if arbitrary, for expressing a concept consistently, and thereby let the potential searcher know what to expect about content of the storage file. Even if the "use" rule should be as absurd as "Zebras" use "ELEPHANTS," the searcher is better off than if "Zebras" does not occur in the alphabet at all. This point is being made so that searchers will not reject out of hand a "use" reference of which they disapprove, or alternatively, will not assume that the hierarchic arrangement can be used without reference to the alphabetic one.

Consider what would happen in this example if the searcher consulted the hierarchy under "mammals" and found "zebras" was not listed. The assumption would probably be made that "mammals" was the logical substitute for "zebras," being the more generic term and also the end of the hierarchic line in trying to locate "zebras." But the "use" reference said that documents about zebras would be found by searching the index term "elephants," and presumably the indexer followed that rule. Therefore, following only the hierarchy and being true to one's own sense of logic can sometimes lead a search astray. This is not a plea to abandon common sense when searching, it is just a warning about the need to be careful and be thoroughly familiar with how to use a thesaurus.

Enough preliminaries. Let us get to a concrete example of how the thesaurus figures into searching, by assuming that we need to retrieve the document that the hypothetical indexer put into the storage system during Chapter 4. We will have to pretend we do not know the title of any particular document in the system or how it was indexed, because even if the indexer and searcher were the same person, memory could not usually be relied upon. However, let us assume we are somewhat experienced searchers of this particular retrieval system, and we have at least moderate knowledge of the terminology of the subject involved in the inquiry, as searchers should. Sometimes searchers quickly acquire familiarity with terminology by consulting a textbook or review article on the subject. With our background we are able to make some inferences and educated guesses about an inquiry and an inquirer, while being cautious enough to double back on our track if actual experience shows any assumption to be of doubtful validity.

Suppose the inquiry reads: "How long can batteries be expected to last when they are used only for radio work on boats cruising in tropical waters?". The first requirement is to thoroughly understand the question. The inquirer used the word "boats," not "ships" or "vessels." This combined with the word "cruising" tells us that the inquirer's interest probably lies in radio work on privately owned yachts rather than on commercial vessels, because a commercial cruising vessel would probably be called a ship (but this is an assumption, so to be safe we had better try to confirm it with the inquirer before proceeding. Assume that he is readily available and confirms it.) A private boat cruising in the tropics could be expected to be equippped for cw (Morse code) communication. If the inquirer is not participating actively in the inquiry and search formulation process, it is best to make a few _alternative_ assumptions, then consult the thesaurus to learn what other strategy-related questions may need answering, and then contact the inquirer before going ahead with the application of the search strategy. Sometimes the inquiry is not for so serious a purpose as the intermediary may think, or sometimes the "real" question will be more fully disclosed once the intermediary can speak knowledgeably about the subject of the search _or_ its likely bibliographic outcome.

We won't reiterate here what we learned about entries in the hypothetical thesaurus in Chapter 4, but let's go through the keywords in the inquiry now and see how we fare. We know that "batteries" is a thesaurus-word, but we have to look it up to learn whether it has narrower terms or other relationships of which we should be aware. "Cells (electrical)" is listed as being narrower than "batteries" and "power supply" is listed as being broader. We reject "power supply" because of its note (chapter 4), but decide that we had better make both "batteries" and "cells (electrical)" search terms. We cannot be sure an indexer would have used both terms for the same document, so we note that we should search for them in a logical OR relationship rather than an AND relationship. "Last" is not in the thesaurus, so we are on our own. How might that be expressed? We look at the entry for "time" and find several terms, including "duration." We might try "duration" in an OR relationship with "time" when we start the matching process. "Radio work" we have only guessed to mean "cw transmission." We don't have to go through the steps the indexer did to arrive at "continuous waves" and "transmission." Remember why? The indexer had the system's supervisor put in a "use" for "cw," and had "continuous waves" added to the related terms note for "transmission," which is where we now encounter it. However, "continuous waves" has "Morse code" among its related terms, so we add that to our shopping list, just in case, with the intent of ORing it with "transmission." We find "boats" and learn that "cruising boats" are distinguished from "racing boats," so we choose both "boats" and "cruising boats" to be sure we will not miss any pertinent references. Tropical waters we wonder about because we find that "tropical" is not in the thesaurus and we don't know whether an indexer might have translated "tropical" into "temperature," "climate," "geographical areas" or what. Boats obviously are found on water (or oceans, or seas, etc.) so it seems unlikely that a term for water will help the search. Because we have quite a few terms to use, specificity of the search is not of concern, so we might decide to skip the redundant concept of "water" for now.

Now we are ready to do the Boolean algebra necessary for formulating the search concretely, using thesaurus terms. The preliminary strategy looks something like this:

batteries or cells (electrical)	AND	time or duration
AND		AND
radio or transmission or Morse Code or continuous waves		boats or cruising boats

Then, if possible, we should check the formulation with the inquirer to learn whether our assumptions are correct, whether we have forgotten anything, whether he wants to see our preliminary results, etc. Additional use of the thesaurus may be necessary in discussing these matters with him. If the search is a manual one, the ANDing and ORing procedures may seem like over-elaborate preparation, but actually they are necessary to either a manual or machine search if it is to be done well. The strategy, per se, may be kept in the searcher's head in the case of a manual search, because human beings have different capabilities than computers, but the strategy must be thought through in the same systematic manner.

Summary of Thesaurus Use During Indexing and Searching

Diagram 1 (Chapter 1) may be more meaningful now that the reader has been considering what happens during indexing and searching. During indexing, document descriptions are readied for the storage file by using the thesaurus as carefully as possible, with the intent of making the document as easy to find as possible. During searching, inquiries are indexed via the thesaurus in much the same way as documents were indexed in the hope of matching index terms in the storage file, to recall pertinent document references, but reject those which are not pertinent. The function of the thesaurus is to make the match between index terms and search terms effective, without sacrificing too much in efficiency (getting too many or too few references).

CHAPTER 6

MAINTAINING A THESAURUS

Background

The language of a thesaurus is a living language, and therefore is always subject to change. It must be clear to the reader by now that a thesaurus can be considered to contain a subset of English, French, or whatever natural language it represents. The reason thesaurus terms are limited to a subset of a full language is because the richness of natural languages allows a concept to be expressed in so many different ways that the game of hide-and-seek (indexing and searching) becomes too challenging for practical purposes. Consider having to formulate a search having to do with furniture on which to lie down, for a retrieval system that allowed all of the following:

> the divan was recovered in ...
> a couch was provided for...
> the reclining place was in use when...
> there were two sofas at...
> Will a cot suffice?
> Has a bed been sent to...
> .
> .
> .

When we speak of natural language in this context, we include the technical language(s) of the native speakers, and these technical languages usually contain a considerable number of expressions or specialized word usages which are not in the general dictionary for that natural language. This is significant because most technical languages are much newer than the overall natural language of which they are a part, and therefore tend to be less "settled." That is, they are subject to a greater rate of change. Examples we have cited are the names of particles in the field of physics, the names of organisms when the field of virology was beginning to develop, terms in the electronics field, and so on.

The Problems of Real Systems

A thesaurus which does not keep up with changes in the language which it embraces, obviously becomes outdated, inaccurate, and unwieldy. However, the other side of the coin is: whenever changes are made in a thesaurus, both it and the storage and retrieval system develop consistency problems. Past indexing and, thus, searching of past indexing poses a dilemma: should past indexing be changed or should the thesaurus and the system built by means of it have stages of existence corresponding to various editions of the thesaurus?

Exploring the alternative answers to that question, we find it would be difficult to reindex documents using older language, because, in them, sufficient discrimination was not made about, for example, subspecies of organisms that may have been (unknowingly) involved, or types of physical matter not differentiated until more recently. Therefore, there is little choice when maintaining a thesaurus over a span of time; like all living things, it and its dependent storage and retrieval system have to have elements of discontinuity while flowing forward through time. These discontinuities are reflected in new editions of the thesaurus and new editions of the system. Naturally, the older editions of the system can be searched, but only in terms of the appropriate edition of the thesaurus. For terms that have undergone change, use of any newer edition will not provide an appropriate match at search time.

New editions are not needed until substantial change has occurred, so actually, the existing e- dition contains changes until the volume of accumu- lated gradual revisions precipitates a new edition. Does this mean that a thesaurus under- goes daily change? Maybe, depending on a number of factors, such as: the fluidity (newness) of the subject areas covered by the thesaurus, the size of the storage and retrieval system (large systems tend to be less flexible) or the sophisti- cation and diversity of interests of the system's clientele (highly sophisticated clients are more sensitive to change in their field of specialty than are non-specialists who may be content with "approximate" answers to their questions).

Persons managing thesauri have to be open to change while being as conservative as possible of the status quo. Every thesaurus change may com- plicate memorized indexing practices as well as storage and retrieval; every resistance to change potentially reduces the utility of the system. The compromise often made is something like the following. Indexers and searchers make a note of new terminology, new usages, etc. encountered in their work. Identification is retained of all documents in which a particular new term was encountered during indexing. After several docu- mented occurrences of the nuance, it is considered for entry in the thesaurus either as a new "use" term or a new "accepted" term. It might be decided that a "use" term should be temporarily penciled into the thesaurus and retroactively applied to the affected documents, pending more occurrences of the nuance; after more occurrences, addition of a new term to the thesaurus is in order, and again, retroactive application of the newest thesaurus rule can be made for the small number of recent documents concerned.

Types of Changes

Thesaurus changes may not always involve ad- dition of terms; there may be deletion of terms, or changes of scope notes, and so on. We know that whenever any one term is changed, it is like- ly to affect other thesaurus entries, because of reciprocal notes. Whenever any change is made, all of its ramifications must be traced and changed throughout the thesaurus. This is one of the times when respect for the system of notes accompanying every thesaurus term is at its highest. An example may help to make this point clearer. Suppose a new term is to be added to this thesaurus. Let's hypothesize that in a particular information science storage and retrieval system there are a great many documents about medical communication, particularly about how to keep physicians up to date in their field. Imagine a new topic, such as "teaching physicians how medical auditing is likely to be done in hospitals in the future." The information system is to help to prepare them for new demands their community hospitals may make when these physicians send patients to the hospital. Half a dozen documents about the subject have arrived, but this thesaurus does not contain the term "auditing," let alone "medical auditing."

The first thing to be done is to examine the concept for meaning, to be certain that "medical auditing" is not simply a new way of expressing a concept already provided for in the thesaurus. Due consideration proves that the concept is not covered. After discussion, a decision is made to add the term "auditing" and then a "use" reference for "medical auditing" which reads, "use: BIOMEDICAL SCIENCES; AUDITING." Now we have to decide where "auditing" fits hierarchically. We find that "budgeting" has "finance" as its more generic term, and "auditing" seems to be on a level with "budgeting," so "finance" has "auditing" added to its listing of narrower terms. While we are looking at "finance," we consider which of its other narrower terms should be listed as related to our new term, "auditing," since its hierarchic siblings are prime candidates for a close semantic relationship. We find "accounting" to be the only one of the siblings that seems closely enough related to be mentioned in the entry for "auditing," but we notice "responsibility" as a related term to "finance" and think it should be listed as a related term to "auditing" too. We wonder about "inventorying" but find that it says: "use accounting," so there is nothing to be done with that. Now, is there anything else to be considered...?

The foregoing is not the easiest type of addition to make to the thesaurus. More trivial examples are:

1) The thesaurus specifies a "use" relationship between a specific term and a more generic term: ex., "Scholarship use awards." You would like to establish scholarships as an accepted term. This is done by making "awards" its broader term instead of its reference term. Any desired notes can be added ad lib.

2) The thesaurus specifies a more-or-less synonymous relationship between two terms: ex., "Senior citizens use older adults." Suppose you have particular reason to separate out a group of older adults to be known as "senior citizens." "Older adults" is not really a broader term in this case, so the hierarchy will reflect "senior citizens" and "older adults" as siblings under "adults."

3) Still more trivial is the case where the thesaurus specifies a "use" rule that needs to be changed: ex., "allocation use planning." You wish to have it read: "allocation use resources." The "refer from" note for "planning" has "allocation" deleted from its membership, and the "refer from" note for "resources" has "allocation" added to its membership.

Given a complete set of terms affected, thesauri are easy to change. For an ongoing system, the difficulty lies in the storage and retrieval system inconsistencies which result from thesaural changes. Changes are made as infrequently as possible, within an overriding policy to change the thesaurus as often as necessary to accommodate changes in the language which it encompasses.

CHAPTER 7

COMPUTER-ASSISTED THESAURUS CONSTRUCTION AND MAINTENANCE

Roles of the Human Compiler and The Computer During Construction

This thesaurus was constructed with the help of computer programs, designed to do most of the clerical and editorial work associated with thesaurus building. The human compiler merely decided what entries would be included in the thesaurus as either "accepted" or "use" terms. "Accepted" terms listed all narrower terms, any related terms that came to mind and any scope notes. "Use" terms specified what term or combination of terms to use instead of them.

After the compiler's decisions were keyed, to make them machine-readable, computer programs did almost everything else:

1) the hierarchy was generated from the narrower term listings, with error messages about any "accepted" terms that,

   a) turned out not to have been assigned a broader term, or

   b) terms that were inadvertently assigned more than one broader term.

2) All reciprocals were generated by computer, that is,

a) "related terms" were made reciprocal and any resulting redundancies were eliminated,

b) "refer from" or else (in the case of more than one "use" term) "partial refer from" notes were generated for every "use" entry, and

c) the "broader term" specification was created from each "narrower term" originally supplied by the human compiler.

3) Any terms referenced in a note through a "use" term, "narrower" term, or "related terms," but not found among the compiler's original keyed entries, was created, and flagged as creations of the computer program. This called attention to any thesaurus omissions (terms referenced but not there). Also any misspellings of "accepted" terms which appeared within the notes of another entry, because they would appear twice, with different spellings. For example, the computer discriminated singulars from plurals and hyphenated from non-hyphenated forms.

4) Human response to errors of omission or commission pointed out by the computer program was necessary, but this type of response was infinitely simpler than attempting to manually create all of the reciprocals and discover all of the logical or spelling errors. With no complaints or censure, the computer kept "doing it again" until the compiler got all the thesaurus entries right--right in the sense that inconsistencies and logical omissions had been resolved, and then

5) the computer and its peripheral equipment, including a somewhat versatile printer, generated the final copy of the alphabetic and hierarchic arrangements, also the single-word index to multi-word terms. The same printer was fed the data for the

title page, table of contents, instructions to users, etc., and the master for this entire book was in page-proof before submission to the publisher. Copies are made from the publisher's photographic reproduction of the master and are distributed in response to orders.

Keeping the Thesaurus Updated with Computer Help

New editions of a thesaurus already in machine-readable form are easy to print on demand, by simply supplying to computer programs for the purpose, all information about changes that are to be made. The type of information supplied can be that specified in the first paragraph of this chapter, if one has access to computer programs with the capability of the ones used to create this book. Under such circumstances the programs will perform one or more iterations of the updating process; the number of iterations required depends on how quickly error-free updating data is supplied to the updating programs. Theoretically, only one pass by the computer is needed, but in practice, the human beings concerned usually make some errors, even if they are only keying errors, and the computer calls them to attention. The computer will continue to call errors to attention until all new entries are complete, and consistent, with previous thesaurus entries.

We know from the foregoing chapter that it is probably desirable to formally make new editions of the thesaurus only when it seems necessary for reason of volume of changes. Recent indexing can be redone to make it consistent with thesaurus changes, up to a point. It is at this point that "cutoff" is declared and the system as a whole moves into a new phase. However, computer updating can be utilized without declaring a new edition in any of a number of ways, such as:

1) distributing "change" instructions for manual insertion into the current "working" thesaurus, with the understanding that the a change is retroactive to the cutoff date of the last edition, or

2) reprinting the thesaurus and flagging all changes, with the same understanding about retroactivity.

Machine-Aided Indexing Via a Computer-Readable Thesaurus

In chapter 4, we considered how a human indexer decides which indexing terms to apply to a document, and found that it was somewhat standard procedure to, among other things, match keywords in the document's title with terms in the thesaurus. If matched with a "use" term, the "use" rule was automatically followed, for the sake of system consistency. A computer program can follow any straightforward rule. When an "accepted" term was matched, the scope note was consulted to determine appropriateness of the match.

Could a computer do this? In some cases, yes; if not, it could at least call the circumstances of indecision to a human indexer's attention. If no match with the thesaurus is possible, a computer program can easily call this to an indexer's attention. If indexing needs to be done in more depth or with more precision than from title words alone, additional text, such as the document's abstract or summary, its section headings, or legends from figures illustrating the document can be fed to the computer for attempted thesaurus match, in the same manner as described for processing the title. The human indexer then participates in whatever decisions the computer program cannot make, given the thesaurus rules above.

On the basis of research, interactive procedures can be expected to produce more dependably systematic indexing procedures than when human beings work without computer assistance. Interactive procedures increase indexing consistency among groups of indexers, and prevent misspelling of terms (every term has to match the thesaurus spelling). Experiments by this author[14], Salton[15], and others[16] have shown this to be quite an effective process. Tests on human indexing, indigenous to commmercial indices, have shown (only hinted at in published articles) human indexing to be of less dependable quality than is processing titles alone by computer, to arrive at thesaurus match, and subsequently at a set of indexing terms for the document.

Why isn't computer-assisted indexing practiced by commercial producers of subject indexes? Start-up costs for it are expensive if the computer programs have to be written and interactive communication equipment provided to each indexer concerned. The process is especially expensive if titles and other data have to be keyed solely for use during indexing. However, most of these costs could be either amortized or circumvented to bring them to acceptable levels. The real impediments seem to be 1) creating an unambiguous thesaurus and getting it into efficiently machine-usable form, and 2) understanding how results of laboratory experiments (Salton's, for example) could be applied to benefit real-world index preparation.

REFERENCES

14. Schultz, Claire K. "Cost-effectiveness as a guide in developing indexing rules." Info. Stor. and Retr. 6:335-340, 1970.

15. Salton, Gerard. Information Storage and Retrieval, A Scientific Report no. lSR-7 to the National Science Foundation from the Computational Laboratory, Harvard Univ. June, 1964.

16. Montgomery, Christine and D. R. Swanson. Machine-like Indexing By People. Am. Doc. 13(4); 359-366, 1962.